Praise for *Self Defence*

'Dan Davis is a world-class scientist and one of the best science communicators of his generation. From the role of tobacco scientists in promoting stress to why astronauts take antihistamines, *Self Defence* will answer every question you have about your immune system and many more. It is deeply absorbing, wise and beautifully written. Packed with astounding science, compelling stories and ideas, it will change the way you think about your body and live your life'
CHRIS VAN TULLEKEN, author
of *Ultra-Processed People*

'The immune system holds the key to modern diseases and this great book from a world expert offers practical tips and dispels many myths on how best to enhance it'
TIM SPECTOR, author of *Food for Life*
and co-founder of ZOE

'Utterly absorbing. Nearly every paragraph brings a revelation'
BILL BRYSON, author of *The Body*

'Expertly cutting through all the myths, marketing hype and simple slogans, this is the best book on immune health I have ever read'
ALICE ROBERTS, author of *Crypt*

'A beautifully written overview of our immune system. But perhaps even more importantly, this book is an instruction manual that guides the reader through the blizzard of popular science and pseudoscience that obscures discussions about our health'
BRIAN COX, physicist, author
and broadcaster

'There is so much misinformation in the health and science arena that it can be really challenging to pick the truth out of all that fiction. *Self Defence* is a perfect antidote to so many of those assumed truths about immune health that have gone unchallenged for way too long. It is a refreshing, expert examination of health myths; beautifully written, packed with science but always accessible. I learned a lot'
SUZANNE O'SULLIVAN, author
of *The Age of Diagnosis*

'The ultimate self-help book – genuinely one of the most worthwhile books I've ever read. Davis is a captivating and authoritative guide, addressing many seemingly simple questions about our immune health, from vitamins to vaccines, from stress to the microbiome. We learn how to learn about the only thing in the universe that's arguably more complex than the human brain'
NICK LANE, author of *Transformer*

'This is the clearest, most authoritative and most honest guide I have seen to the kinds of questions about our health and our immunity that we encounter all the time. What is the best diet? Does exercise help? What does our microbiome really do? Is good sleep so important? With patience, humility, humanity and generosity, Dan Davis provides a primer on immunity that should be on everyone's bookshelf'
PHILIP BALL, author of *How Life Works*

'An engaging and thorough look at the factors affecting our immune health. If you've ever been tempted to take steps to "boost your immunity", read this first!'
SARAH GILBERT, author of *Vaxxers*

'Professor Davis unravels one of the most complex and vital elements of human biology: our immune system. Disentangling the science of immunity and addressing the key questions we all want answered, this book is a masterclass in science communication and essential reading for anybody who wants to understand their health'
RUSSELL FOSTER, author of *Life Time*

'A fascinating guide to the immune system, cutting through the myths with clear, evidence-based science. Daniel M. Davis empowers readers to make informed choices about their health in a world of so much misinformation'
RHIANNON LAMBERT, author of *The Science of Nutrition* and co-host of *The Wellbeing Scoop*

'Admirably clear, readable and reliable. Davis writes with infectious enthusiasm about the beauty and complexity of human immunity'
GAVIN FRANCIS, author of *Recovery*

'Wellness gets the expert treatment as Davis filters out the bullshit, hype and hearsay in a clarifying sweep to reveal the fascinating complexity around how our body maintains health, and how we can help it'
GAIA VINCE, author of *Nomad Century*

'Read this book! Your immune system is notoriously confusing yet vital and thus worth understanding. Fortunately, this succinct, clear, insightful and fascinating book from one of the world's experts on immunology will answer your most important questions about how best to help your body defend itself and stay healthy'
DANIEL LIEBERMAN, author of *Exercised*

'The body is complicated: you need a book, not a slogan, to unravel what helps keep it healthy. Fortunately, Davis has written an excellent and expert guide to navigate us through the minefield of theories and speculations about what might boost our immune system'
MARCUS DU SAUTOY, author of *Around the World in 80 Games*

'We all retain some traditional beliefs about what lifestyles and everyday remedies are good for us; this clearly written book by a distinguished expert tells us which ideas are grounded in science, and which are not. An enjoyable read with messages that can improve our health'
MARTIN REES, author of *If Science is to Save Us*

'A brilliantly crafted immunisation against the fads, fixes and fraudsters that seek to improve our immune health. Davis draws on the past, present and future of immunology to inform and explain what we can and, importantly, cannot do. Ultimately what he shows is that the immune system is wondrously complex and that simple soundbites, however persuasive, will have little real impact'
JOHN S. TREGONING, author
of *Live Forever?*

'In *Self Defence*, immunologist Dan Davis explains everything that we know about our immune systems with clear, concise and engaging stories based in the latest science, not hype. So, if you're tired of all the people trying to sell you things that they swear will improve your immune function, but then never do, this book is for you. Dan walks us through the often complicated answers to all our questions and shows us that there is no such thing as a one-size-fits-all solution to immune health'
THERESA MacPHAIL, author of *Allergic*

'*Self Defence* uncovers how science can help us truly understand and optimise our immune system – free from hype and misconceptions. Dive in to see how thoughtful, evidence-based exploration can lead to better health, resilience and self-awareness'
ROGER HIGHFIELD, Science Director,
Science Museum and co-author
of *The Dance of Life*

'Why does weight-loss drug Ozempic reduce the risk of dementia, and what's the connection between the bacteria living in your gut and the risk of allergies? The answer – as in so many parts of biology – is the immune system. *Self Defence* takes us on a fascinating tour of our bodies, linking happenings inside our cells to everyday practical advice, and dealing with pervasive myths about vitamins, inflammation and more along the way. This book will do more for your health than any bottle of vitamin C tablets!'
ANDREW STEELE, author of *Ageless*

'A masterful exploration of what we know (and don't know) about the workings of this most crucial of bodily functions. In a world of sales patter, soundbites and social media videos, scientific rigour, clarity and nuance is much needed. *Self Defence* provides precisely that. Davis dissects claims and widely held beliefs surrounding the workings of our immune system, and shines a light on what is myth, hype, uncertainty and scientific truth'
GUY LESCHZINER, author of
Seven Deadly Sins

Books by Daniel M. Davis

The Compatibility Gene
The Beautiful Cure
The Secret Body
Self Defence

Self Defence
A Myth-busting Guide to Immune Health

DANIEL M. DAVIS

THE BODLEY HEAD
LONDON

1 3 5 7 9 10 8 6 4 2

The Bodley Head, an imprint of Vintage,
is part of the Penguin Random House group of companies

Vintage, Penguin Random House UK, One Embassy Gardens,
8 Viaduct Gardens, London SW11 7BW

penguin.co.uk/vintage
global.penguinrandomhouse.com

First published by The Bodley Head in 2025

Copyright © Daniel M. Davis

The moral right of the author has been asserted

The information and suggestions contained in this book are not intended to replace the services of your physician or caregiver. Because each person and each medical situation is unique, you should consult your own physician to get answers to your personal questions, to evaluate any symptoms you may have, or to receive suggestions on appropriate medications. The author has attempted to make this book as accurate and current as possible, but it may nevertheless contain errors, omissions or material that is out of date at the time you read it. Neither the author nor the publisher have any legal responsibility or liability for errors, omissions, outdated material or the reader's application of the medical information or advice contained in this book.

Excerpt from 'Learning to Live with the One Body You've Got'
by Beatrice Garland © Templar Poetry

No part of this book may be used or reproduced in any manner for the purpose of training artificial intelligence technologies or systems. In accordance with Article 4(3) of the DSM Directive 2019/790, Penguin Random House expressly reserves this work from the text and data mining exception.

Typeset in 13.5/16 pt Garamond MT Std by Jouve (UK), Milton Keynes
Printed and bound in Great Britain by Clays Ltd, Elcograf S.p.A.

The authorised representative in the EEA is Penguin Random House Ireland,
Morrison Chambers, 32 Nassau Street, Dublin D02 YH68

A CIP catalogue record for this book is available from the British Library

HB ISBN 9781847927569
TPB ISBN 9781847927576

Penguin Random House is committed to a sustainable future
for our business, our readers and our planet. This book is made
from Forest Stewardship Council® certified paper.

For Marilyn Davis and in memory of Gerald Davis

Contents

Introduction: How we differ   1

1. Orange Juice and Sunshine – or, do vitamins help immune health?   17
2. The Aliens Living Inside You – or, does the microbiome affect immune health?   34
3. The Evidence of Weight – or, does weight affect immune health?   58
4. The Yin and Yang of Exercise – or, how much exercise is best?   77
5. A Reason for Calm – or, does stress impact immune health?   97
6. Inner Beauty Sleep – or, does sleep help immune health?   115
7. The Two Big Systems – or, does immune health affect mental health?   133
8. 100-Year Long Immunity – or, how does immune health change as we age?   152
9. In the Pipeline – or, what big new ideas are on the horizon?   173

Conclusion: the overarching journey   195

CONTENTS

Acknowledgements 203

Glossary 205

Further Reading 211

Notes 215

Index 277

> We need each other:
> we could live as friends.
> If you'll forgive me
> I will make amends.

'Learning to live with the one body you've got',
Beatrice Garland

Introduction: How we differ

Every year, as a professor of immunology, I give talks to three different audiences: other scientists, students and the public. My talks to other scientists will be greeted with a ripple of applause, and then some of them will stand up to challenge me, as they should, on the details of my team's recent experiments. To almost anyone outside the room, their questions would be impenetrable, couched as they are in a specialised vocabulary of names and abbreviations, necessary to communicate nuances. (Did you check if the TCR is triggered? Great question; we're studying that now.) After a while you get used to the jamboree of C-3PO- and R2-D2-like jargon.

At my lectures to students there isn't usually much applause, but their enthusiasm shows in other ways. Sometimes I'll be cornered, or emailed later, to clarify some details (or check what's examinable). It's after a talk to the general public that the applause is the most generous. Then follows – every time – a series of seemingly simple but actually really difficult questions, mostly, if not exclusively, about how to achieve good immune health. Is it true that orange juice can help ward off colds? How does age effect our immune system? Is mental health really connected to inflammation? How important is vitamin D?

During a brief Q&A there isn't time to answer these questions properly, because more than a snappy sentence or two is required. But these big, knotty issues are what I want to tackle in this book. To answer such questions, and others like them, I want to take a deep dive into the immune system in all its complexity, and the vital role it plays in illnesses – pretty much all of them. This will, I hope, enable you to make more informed lifestyle choices to improve your own immune health. We'll learn that some things are proven to help, others will only help certain people, and a lot of things – while not without *some* level of scientific backing – are much more speculative than the advertising slogans would have you believe.

One thing that is abundantly clear is that good immune health is vital. Without it we succumb to any number of different infections and illnesses. Children born with defects in their immune system, such as lacking a certain type of immune cell, tend to suffer repeated and severe infections or develop autoimmune conditions. In its rarest and most extreme form, immunodeficiency can be life-limiting and life-threatening. Take, for example, the tragic situation of David Vetter, famously known as the boy in the bubble. Born in 1971, with a severe problem in his immune system, he spent his whole life in hospital in Texas, in a special isolated environment. Though he remained relatively free from exposure to germs, sadly he died at the age of twelve from cancer caused by a virus.

Another thing we can be certain of is that immune health is complicated. Partly that's because our immune system must be able to fight off different kinds of germs, including bacteria, viruses and fungi, each of which has to be detected and destroyed in lots of ways so that it is hard for germs to evade the attack. On top of this, we also need the immune system not to set about attacking any of the body's healthy tissues and cells.

This is what my research team and I are immersed every

day in trying to understand. For example, we examine how an immune cell will latch onto a cancerous cell, recognise it is dangerous and send over toxic molecules to kill it. The different kinds of immune cells we study also include those that can target a dangerous bacterium, engulf it and destroy it. It all feels a long way from everyday experience, even other-worldly, but such processes are profoundly real and happening inside my body and yours right now.

A simple way of putting it would be that the immune system fights off anything which is not part of the human body. But on reflection this can't be true – or at least not the whole truth. Granted, germs are not part of the human body, and dangerous germs should be killed. But food is not part of the human body either, and we don't want the immune system to react against harmless food. Plenty of germs are harmless, too. In fact, there are billions of bacteria living in your gut that are actually beneficial, and we don't want the immune system to react against normal, resident gut bacteria.

So what do we mean by good immune health? It's a myth to think that *boosting* the activity of our immune system is always a good thing to do. If the immune system is boosted too much, the body's normal healthy cells and tissues can be attacked. Indeed, some diseases are themselves caused by an unwanted immune response against the body's normal cells. For example, the autoimmune disease multiple sclerosis arises when the immune system inadvertently attacks the outer cover of nerves, causing problems in how the brain sends messages to the rest of the body. Other autoimmune diseases include rheumatoid arthritis, systemic lupus and type I diabetes. In all of these the immune system mistakenly reacts against the body's healthy cells and tissues. To tackle autoimmune diseases, we want to *dial down* the activity of the immune system – but not too much, or we risk the immune system no longer responding well to an actual threat, say from a viral or bacterial infection. In other words, good immune health is about having the human body respond to a genuine threat, with precision and to the right extent, with minimal collateral damage. Already that sounds a lot harder to achieve than just *boosting* immunity.

Which is why we need this book. I want to explain what we really know about immune health, and what you can do with this information. My research team and I have been digging away for more than two decades to help show how immune cells recognise the presence of diseased cells in the body, and how different types of immune cell communicate with one another. And we are just one research team: around the world, there are thousands more doing incredible work. From this vast human endeavour, sufficient knowledge has built up that we can now begin to understand what really affects immune health. It's only with this profound understanding that we can make more informed choices in how we live.

I will cover the topics I am most often asked about: vitamins, the microbiome, weight, exercise, stress, sleep, mental health, age and medication. Every one is a crucial aspect of health and well-being in popular culture, and in our daily lives. But every one is also a scientifically complex field, where amid a sea of tweets and hashtags, nuance struggles to surface. That big ideas are often boiled down to a simple hook or headline plays on something innately human: our brains seem hardwired to relish black-and-white thinking. Whenever you read or hear a new headline, it feels as if you've just found out a new secret about how things are, and

INTRODUCTION: HOW WE DIFFER

should take action accordingly to try and improve your immune health.

Yet all too often there are complications underneath. This is unsatisfying, because we are left not quite knowing. By working through the evidence for and against, however, detail by detail, as I want to do in this book, we learn how to learn. In the face of intricacy, nuance and subtlety we'll become sceptical of any unitary, dogmatic approach to improving immune health. As Darwin once wrote, 'Ignorance more frequently begets confidence than does knowledge,' which is exactly what we will see with immune health: scientific knowledge is what we need to make the right choices in our lives, but that doesn't mean the answers are always declarative, prescriptive or simple.

Indeed, whenever you're considering advice or information about immune health, there are three things to be especially wary of: hype, correlations and prejudice. Let's unpack each briefly.

First, hype. Global corporations thrive on selling foods, drinks and supplements that claim to support health, which means billions of dollars are riding on your decision as to what works. One UK high-street chain has 276 products listed under 'immune support supplements'. There's a lot of marketing built around these capsules, pills, drinks, cereals, snack bars and yoghurts, all of which are promoted as boosting immune health. Crucially, these products tend to be regulated under food law, which means there are rules about safety that must be complied with, such as the maximum amounts of particular vitamins and minerals that can be used, but they don't have to be scientifically proven to support immune health. A certain scepticism about any simplistic slogan, especially when it's marketing hype, is thus very useful for good decisions about health. Whenever you read something online or in the media about immune health, as well as what it says, note who's saying it.

Second, correlations. One way of guiding us to what helps us be healthy is to compare how different people live and what they eat, perhaps in different parts of the world, with the types of illnesses they are more or less likely to succumb to. But there's a big problem with this: it's extremely hard, if not impossible, to isolate the effects of any single factor, and things are often linked because of something else. For example, if eating ice cream were linked to skin cancer, this would sound alarming, but it could be accounted for if people who ate more ice cream tended to live in sunny countries and thus were more exposed to the damaging effects of UV light on their skin. To take a more complex example, older adults are more susceptible to autoimmune diseases, but it's hard to disentangle how much of this might be related to ageing itself from the reality that as we age, sleep, diet and exercise patterns also change. To understand immune health, we need more than correlations: for a link to be affirmed, there must be other lines of evidence for it – for example, understanding, in molecular-scale detail, exactly *how* ageing or sleep, diet and exercise, can affect immune health.

Third, your own prejudice. As Einstein once said, common sense is nothing more than a deposit of prejudices laid down in the mind prior to the age of eighteen. We must guard against simple connections that seem to 'make sense' but are really just myths we've grown up with or that pervade our culture. I was brought up with the idea that drinking orange juice, which is high in vitamin C, will help you get over a cold, or stop you catching one in the first place. For a long while, this seemed to 'make sense' to me; after all, taking things as fact, whether from parents, carers or teachers, is instinctual, and almost certainly something we've evolved to do to help us survive. But I now question whether this whole thing about vitamin C is actually true (spoiler: it's not). We must be suspicious of our prejudices. Only by delving into the details and being prepared to embrace complexity can we begin to understand what really helps us stay healthy.

We've been accumulating this wealth of understanding about immune health for centuries, ever since the first vaccination was performed in May 1796, when Edward Jenner took pus from a dairymaid infected with cow pox and inoculated his gardener's eight-year-old son to protect him from smallpox. Since then, we've learnt an enormous amount about how the body can fight off dangerous viruses, bacteria or other germs. We've also learnt that some diseases are not in fact caused by germs. For example, cancer arises when abnormal cells divide in an uncontrolled way. Sometimes this is caused by a virus, but far more commonly it happens by chance, or is triggered by excessive exposure to sunlight, chemicals in tobacco smoke or other carcinogens. Meaning that the immune system must also deal with other dangers to the body as well as infectious germs.

Building on these fundamentals has enabled us to understand how various lifestyle factors influence our immune health. Long-term chronic stress, for example. Stress of any kind, from taking exams to having relationship problems or losing a job, causes the adrenal glands, situated on top of our kidneys, to pump out hormones, one of which is called cortisol. It's natural for levels of cortisol in a person's blood to vary through the day, but when we're stressed our cortisol levels change dramatically to prepare the body for a 'fight-or-flight' response. Such increased levels of cortisol reduce the efficiency with which immune cells engulf germs or kill diseased cells – which is fine for a little while, but if stress persists, our immune system may stay weakened.

From knowledge like this, we can move beyond correlations, prejudices and hype to examine the underlying biological processes and mechanisms that play a crucial role in our immune health, and address precisely how lifestyle factors can affect the immune system. Right now, indeed, our understanding of immune health is burgeoning. The Covid-19 pandemic took us on a long journey of understanding with this virus, and especially our immune response to it. Sequencing the genes of nearly 7,500 patients who needed treatment in intensive care, for example, compared to over 48,000 people who didn't, found several genetic variations in people to be associated with critical illness, many of them part of our immune system. We have learnt that a mild case of the disease correlates with an immune response being triggered early, when the virus first enters the body. A picture is emerging in which our immune response plays a crucial role in determining the symptoms we experience when contracting the virus. Since each of us now will have been exposed to different combinations of natural infections and specific vaccinations, this adds another layer in how everyone's immune system is configured slightly differently for protection against future versions of the virus.

Which brings us to an important theme: we are not all the same. Of course, people vary in all sorts of ways. How we walk and talk. Our political and moral values. The things that make us laugh. How we smile. But under the skin, and even within skin itself, there's a whole other dimension to what we are: a multitude of cells and molecules that make up the body's systems and sub-systems that can be identified only with scientific instruments and analysis. And there are differences between us here too, especially in the immune system. In fact, your immune system is, in some ways, the single most unique thing about you.

Which means that the very notion of 'good immune health' is also something of a myth, at least in the sense that there's an optimum state of immune health we could all attain, so that we would all get over a cold extremely quickly, or not even catch it in the first place. This cannot happen because we are all different. It's inevitable that some people are going to be more susceptible to this season's cold viruses or feel its effects more than others. Indeed, we vary in how susceptible we are to *every* kind of illness, and in the symptoms we experience with *every* type of infection.

Of course, you'll already be aware that some of us get milder symptoms than others when infected with the exact same virus. Symptoms vary for each of us, for lots of reasons: our lifestyle, previous illnesses or age, for example, but also because of something fundamentally important which you can't alter – your genes. Everyone has a very similar genetic inheritance – the human genome – but there are variations in the specific set of genes each of us has inherited and, surprisingly, the genes which vary the most from one person to the next are those in our immune system.

This variation in each person's immune system couldn't be more important for making sense of our immune health. Everything this book will cover, from stress to nutrition to ageing, affects each of us in different ways. There are some general truths, as we will see. Certain lifestyle choices might be generally bad, others beneficial. But some things are likely to benefit you more than me. Others will be especially bad for me, and not you. In other words, something which works for one person — even something which may be very generally true, on average, across the whole population — may still not work for you. Everything is personal. This is the backdrop to everything else: there is no 'one-size-fits-all' path to immune health.

Just as we respond differently to infections, we also vary in how we respond to medicines and treatments. Take, for example, immune therapies, which harness the power of the immune system to fight disease, such as those used in treating cancer. This type of medicine can save the lives of some cancer patients but do nothing to help others. For some especially unlucky patients, the very same therapy can actually trigger bad side-effects, causing the symptoms of an autoimmune disease. Crucially, it's not obvious at the outset what will happen to any one of us individually.

Sometimes personal information can be used to mitigate the chance of something bad happening. A familiar example is how genetic variations correlate with an increased risk of cancer. This kind of information famously led the actor Angelina Jolie in 2013 to have both her breasts surgically removed, and subsequently her ovaries and fallopian tubes. A genetic test had established that she had inherited a mutation in a particular gene known as BRCA1 which, together with her family history, meant that she had an 87 per cent chance of developing cancer herself, if she hadn't had surgery.[1] Since then, a huge global endeavour has continued to investigate relationships between genes and diseases.

To get a sense of the scale of this, biobanks in the UK and Finland each hold genetic information for around half a million volunteers, and an analysis in 2022 uncovered nearly a thousand links between genes and various diseases, from cancer to sudden infant death syndrome.[2] Most of these

INTRODUCTION: HOW WE DIFFER

genes won't have a big effect on anybody's risk, because most gene variations only have a small impact. But what's clear now is that pretty much every one of us is susceptible to some disease or other to some extent, on account of our genetic inheritance. In other words, we are all sub-optimal. Or we're all special, depending how you look at it. And something else emerges: something very surprising the first time you come across it.

Most genes affect how we fare with one type of illness, such as the fact that BRCA1 directly impacts a person's risk of developing cancer. But our susceptibility and our resistance, and the symptoms we experience, for pretty much *every type of illness*, from multiple sclerosis to rheumatoid arthritis to cancer to schizophrenia, are affected by the versions of immune system genes each of us has.

Let's explore a specific example. Someone who has inherited a particular immune system gene called HLA-B27 is about 300 times more likely to develop the autoimmune disease ankylosing spondylitis, which causes back pain, stiffness and tiredness.[3] Inheriting this gene does not make contracting the disease inevitable; in fact, the incidence is rare, even among people who have the gene. Around 8 per cent of people in the UK have it, and the vast majority do not suffer from this disease.[4] Nevertheless, most people with ankylosing spondylitis have inherited this gene.

This would seem to imply that inheriting this gene is a bad thing. But there's a flip side. Inheritance of this very same gene also seems beneficial in fighting HIV. About 1 in 300 people infected with HIV have the capacity to limit the virus's ability to multiply, so they don't go on to develop AIDS for a very long time, and among them the immune system gene HLA-B27 occurs frequently. In other words, a particular inherited gene can make you more susceptible to one type of illness, but actually be beneficial against another.

Any notion of a genetic hierarchy – the sense that some people might be inherently better or stronger than others – is obviously dangerous. Indeed, such thinking has throughout history led to some of most horrific crimes against humanity. Rather what makes the variation in our immune system genes so wondrous and meaningful is precisely that there's no such thing as an ideal set of immune system genes: one genetic inheritance can be beneficial in one respect but worse in another.

Because we each inherit several versions of immune system genes, there is a further and even more subtle level of diversity within a single person's body, making every one of us individually stronger. For me this is one of the most profoundly important things we have learnt in all human biology: a yin and yang within everyone.

Our genetic differences also play a role in how well we respond to vaccines. For example, responses to the Oxford-AstraZeneca vaccine for Covid-19 vary according to which versions of immune system genes a person has.[5] The vaccine still protects everyone, but there can be variation in how well each of us is protected, or how long for. Given this heritable component to how well a vaccine works, it's possible that in the future, vaccines could be tailored to match a person's genes, or genetic analysis might indicate who should be prioritised for booster vaccines. We'll explore the revolutionary possibility of personalised vaccines further in the final chapter.

Unquestionably, too, there are sex differences in immune health. Almost all autoimmune diseases, for example, are more frequent in women, with multiple sclerosis being twice or three times as common.[6] There are also general differences in how men and women respond to infections. Again, take Covid-19: men are more likely to die from it, whereas women are more likely to suffer from long Covid.[7] Genes are one reason, because an important immune system gene is positioned on the X chromosome, so women, who have two X chromosomes, effectively have a double dose of it. This gene is involved in fighting certain viruses including the flu virus, which is one reason why women tend to respond more strongly when some types of viruses are in their blood.[8]

Hormones probably also play a role, though confirming this is difficult because sex-based differences could arise out of any number of social, economic or cultural factors. Evidence that hormones are directly important, however, comes from the analysis of blood samples from trans men undergoing testosterone therapy.[9] Changes in their immune system were apparent within months.[10] What this means for the immune health of trans men is not clear, but it is good evidence that sex hormones impact immune health. This isn't, and can't be, an immediate prompt for action, but it is possible that in the future specific medicines will be developed that work better for women or men.

All this genetic diversity means that your immune system has a uniqueness to it even at birth, before anything much has happened to you, before your immune system has fully developed, irrespective of whether it has ever been deployed to defend you against any type of disease, and irrespective of any specific lifestyle choices you have made. From this come two big messages. First, your immune system is fundamentally different from mine. The other thing is: next time you are ill, perhaps in bed with a fever or a sore head or just feeling exhausted, and others around you who have caught the exact same infection seem less ill or have bounced back more quickly, it does not necessarily mean you are somehow weaker, too highly stressed or low in some vitamin or other. Modern culture, with its drive for simple answers, the health industry's relentless messaging, and the lucrative business of offering solutions, perhaps creates a particular danger: that we'll blame ourselves for how our body is reacting, or why we are ill or seem to be suffering more than others. But thinking you didn't look after yourself is not going to be the sole reason. It might just be your basic nature, such as your genetic inheritance, which makes you especially impacted. And it is very likely that you'll be stronger in some other situations, such as dealing with a different type of infection.

In Jack Kirby and Stan Lee's X-Men comic books first published in 1963, which are essentially stories about how we differ, mutations have given some people superpowers. Now, variant genes can't give someone an ability to fly, but a real-life genetic superpower is that every one of us is equipped with an ability to fight off certain diseases better than average. We are all mutants, if you like, each with our own immune superpower.

This understanding of our individuality and how this intersects with disease and illness couldn't be more important for society too. Long ago, humankind made up reasons for illness that had devastating consequences. For example, when the Black Death arrived in Europe in 1347 it was common to believe that humanity was being punished by God, which in turn twisted into a desire to kill enemies of Christ. Jews and other non-Christians were accused of poisoning water wells as an attack against Christianity. Under torture they confessed. In vengeance, thousands were murdered in France, Austria and Germany. This then helped seed the following century's Spanish Inquisition. So a comprehensive misunderstanding of disease played its part in forced religious conversion and burning people at the stake. The great irony is that human diversity, and the very fact that immune system genes are so varied, is central to how we survive disease.

The bottom line is that much of our immune health is simply down to the fabric of who we are – unique and unchangeable for each of us individually – but that's not to say we can't work with what we've got. Identical twins carry the exact same genes but vary enormously in their own health. One reason is that, as we grow up, our immune system is shaped by our lifestyle, exposure to germs, and more. Much as babies' faces tend to vary, but not as much as adult faces, our immune systems become even more diverse, reflecting everything we've been through. What we do matters.

Plainly, the human immune system is not a simple part of our anatomy; it is a multi-layered, dynamic lattice of interlocking genes, proteins and cells, amounting to one of the most complex and thrilling realms of modern science. From what we now know, some simple truths emerge, such as the negative impact of long-term chronic stress as well as the positive benefits of getting a good night's sleep. Yet other ideas, such as the ability of vitamin C to stop a cold, turn out to be more myth than fact.

All of this, the science of the immune system – immunology – is a subject whose time has come. Almost every day we are presented with new theories about immune health, new advice or new kinds of drugs and treatments. In fact, it is no exaggeration to say we are at the beginning of a revolutionary time for immune health. An enormous amount of basic knowledge is in place, and we are now learning to apply this knowledge practically, for living more healthy lives.

We don't yet understand everything. Many questions I commonly get asked remain very hard to tackle. How does burnout impact immune health? How do ultra-processed foods affect the immune system? Is there any truth in the idea of man flu? Here the science is still ongoing. But ever since Jenner first vaccinated a young boy more than 220 years ago, there's been an enormous global endeavour to understand how the immune system really works. By journeying through the science – by carefully and honestly answering some key questions and exposing some long-held, yet harmful beliefs – I hope you will come away able to make more informed choices about all manner of ideas and products for living healthier, happier and stronger.

I

Orange Juice and Sunshine – or, do vitamins help immune health?

Nutrition and immune health are enormous subjects. Plenty of books and magazine articles list foods or supplements purported to help immune health. It's not that everything in such lists is a myth, but it is extremely hard to tell. Of course, companies have a bias when they tell us about their own products. But more subtly, every new idea is entangled with personalities and opinions and agendas. When experts offer advice, we must know if they are presenting the consensus view based on the available data, or a fringe opinion they happen to hold. This is what allows us to make informed and empowered decisions about our own immune health – especially in today's high-tech, polarised world.

Fads come and go. This chapter cannot cover every possible nutritional advice for immune health that you could come across, when a new idea is just around the corner. But there is one story about nutrition and immune health that helps us be informed and forearmed about anything else, because it is a controversy we have been thinking about and studying for over half a century already. A lot of information has accumulated, and the truth is that we have been misled for decades. Vitamin C and colds.

Why do we think vitamin C can cure a cold?

Most animals, including sharks, cats and dogs, produce their own vitamin C. But about 60 million years ago, our ancestors lost this ability on account of one gene becoming inactivated. So all the vitamin C we need must come from food or drink, especially vegetables and citrus fruits and, famously, orange juice. What's more, not only can't we produce vitamin C, but we also can't store it. So we must consume it regularly.

There is no question that vitamin C is important to health. It is a strong antioxidant which means it neutralises the activity of unstable atoms or molecules, called free radicals, that arise from pollutants, toxins and everyday body processes.[1] Vitamin C also helps the body take up iron from certain types of food, such as beans and grains, by transforming the iron into a different form.[2] What's more, vitamin C is involved in the production of many proteins and chemicals vital to the human body, such as collagen, which is essential for tissue healing and a key component of bones, teeth, cartilage, tendons, skin, blood vessels and heart valves. But does it also cure a cold?

For over fifty years, the idea that vitamin C can cure a cold has had a stranglehold on the popular imagination. This is thanks to Linus Pauling, a double-Nobel Prize-winning scientist based in the US. Without Pauling's public advocacy and headstrong attitude, we would not think of vitamin C as we do today. Arguably, the whole multi-vitamin marketplace and the idea of supplements boosting our immune health would look completely different without him.

Pauling was first and foremost a chemist, and is often said

to be one of the greatest chemists of the twentieth century. After his father died when he was nine, he found solace in books and devoted himself to science, against his mother's wishes, often reading scientific papers to look for problems he could think about and solve. By the age of thirty he had published over fifty scientific papers. In mid-February 1931, around the same time as his second son was born, he finished a landmark paper, 'The nature of the chemical bond'. It was published without the normal scrutiny because the journal's editor decided there wasn't anyone else in the world with enough expertise to critically evaluate it.[3]

Needless to say for somebody who won two Nobel prizes, Pauling always worked hard.[4] He had a knack for thinking through issues clearly and working out the overarching rules and principles: 'I like to take a very complicated subject where there is no order ... and think about it for a long enough period that I can find some way of introducing order into it.'[5] But already there lies a problem. Simple, precise and exact rules work exceptionally well for making sense of isolated chemical reactions, but not so easily for the entire human body, let alone something as vast as immune health and nutrition. There are just too many variables.

The onset of fascism from the 1930s was a dark time in Europe. Hitherto, broadly speaking, most scientists had tended to be apolitical, at least in public, on the understanding that the pursuit of knowledge should be kept pure and impartial. But from the 1930s many scientists became more politically engaged, as it became increasingly clear that science affects all sorts of social and economic developments. The trend was probably accelerated by many Jewish scientists fleeing Germany for labs and research centres across the US with shocking and horrifying stories to tell.[6]

SELF DEFENCE

In 1940, Linus Pauling began making political speeches against fascism and arguing that the US should prepare to enter the war; years later, when he read in a newspaper that Hiroshima had been destroyed by the atomic bomb, he called it the 'ultimate immorality'.[7] His stature continued to grow. His textbook for undergraduates, *General Chemistry*, sold well and made him wealthy. All this is relevant to the story of vitamin C because it is about how Pauling became famous.

Many issues of great public importance were still not well understood scientifically. There was much debate over the effects of radiation on the human body, how were radioactive particles spread by the wind, what happens when they fall on the ground, what happens to cows that eat contaminated grass. The available data was vague, and could be twisted to support differing views about the consequences of atomic bombs. Pauling and many other scientists became respected and important public figures, debating these difficult and complex issues in the open, and occasionally going beyond their scientific expertise, enjoying presenting their own opinions. In 1958, he published a book provocatively entitled *No More War!*[8] By this time, he had already won a Nobel prize for Chemistry for his research in understanding chemical bonds. In 1962, he won the Nobel prize for Peace for his activism against nuclear bombs. So by the time he turned to biology and human health, and specifically vitamin C, the public were primed to listen to him.

In 1965, Pauling read a book about high doses of the B vitamin niacin helping people with schizophrenia (an idea not supported today). Something about this struck him as particularly strange. With most drugs, there's a limit to how much can be safely taken. Too much aspirin or paracetamol can be fatal, for example. A new idea germinated in his mind.

Could there be something special about vitamins which made them safe and effective in large doses?

Pauling was in contact with the scientists Albert Szent-Györgyi, renowned for his discovery of vitamin C, and Irwin Stone, who helped develop processes for using vitamin C in food preservation that are still used today, when it is often listed on labels as ascorbic acid. Stone claimed that extremely high doses of vitamin C helped him and his wife heal rapidly after a car crash. He argued that vitamin C is nowhere near as abundant as it used to be in our diet, and so we need to consume more to reap the benefits. These were notions drawn from personal experience rather than carefully investigated science, but Pauling latched onto them all the same.[9]

Certainly it is very hard to know how much of any molecule is needed for the human body to function well.[10] What's more, together Pauling, Stone and Szent-Györgyi realised that the intake of vitamin C recommended at the time was set according to the level at which it prevents the disease scurvy. This, they argued, was not necessarily what might be optimal. So Pauling and his wife began taking very high doses themselves: 3,000 mg a day, and came to feel this mega-dose gave them extra energy and stopped them catching colds.[11] For comparison, the current recommended level in the US is 75 mg per day for women and 90 mg for men, and in the UK it's 40 mg for everyone.[12] (The level varies from country to country depending on the criteria: some still go by the amount necessary to prevent scurvy; others require a certain level to be detected in the blood.) In any case, Pauling and his wife were taking about thirty times any country's recommended amount.

Pauling scoured scientific journals, cherry-picking results which fitted his narrative. In 1970, his best-selling book was published: *Vitamin C and the Common Cold*, in which he claimed that everyone's health could be improved by high levels of vitamin C.[13] Timing helped. The counterculture of the late 1960s had already put a spotlight on whole foods, natural remedies and treatments.[14] Sales of vitamin C surged. New factories were built to keep up with demand, despite many healthcare and scientific professionals maintaining that the evidence was not there, and no one could say how vitamin C helped against colds.[15] Pauling countered by criticising the whole pharmaceutical industry for seeking to uphold the status quo because their own cold remedies were so lucrative.

So what is the truth? A so-called systematic review collects the data available from *all* trials which fit pre-specified criteria. In 2013, a search for all placebo-controlled trials testing the effect of high levels of vitamin C on colds uncovered 29 trials involving over 11,000 people. The conclusion from all this data was that, for the general population, vitamin C does not reduce the chance of catching a cold.[16]

However, for reasons unknown, people taking regular supplements of vitamin C did experience cold symptoms for slightly less time. To be precise, regular supplementation of daily vitamin C reduced the duration of a cold by 8 per cent in adults and 14 per cent in children. In other words, for a cold lasting a few days, someone taking regular vitamin C supplements might feel better a few hours earlier.

The benefits could be greater for particular people. For example, five trials with a total of 598 participants found that athletes who exercised heavily, including marathon runners and skiers, benefitted more from vitamin C supplements. It even lowered their risk of catching a cold. Overall, the scientists running this systematic review of all trials concluded: 'in our opinion, this level of benefit does not justify long-term supplementation in its own right ... [but] it may be worthwhile for common cold patients to test on an individual basis whether therapeutic vitamin C is beneficial for them.'

Future research may more clearly establish specific situations or groups of people for which vitamin C is important for immune health. But in the meantime, many of us simply *believe* vitamin C protects against colds far more than it really does, unaware that this is largely down to one man's crusade. All these years after Pauling's 1970 book, we still tout the idea of orange juice helping us fight off colds without giving much thought to its credentials or veracity.

Can vitamin C help with cancer or HIV?

In 1973, Linus Pauling proclaimed that vitamin C helps with cancer too.[17] A couple of years earlier, the Scottish physician Ewan Cameron had written to tell him that in his small hospital just outside Glasgow, he had achieved great results from giving cancer patients high doses of vitamin C. The medical establishment didn't consider Cameron's study careful enough but, with Pauling now involved, the public got to hear about it on BBC Radio 4, in the *New York Times* and in *New Scientist* magazine.[18]

Following surgery for her stomach cancer, Pauling's own wife, Ava Helen, took high doses of vitamin C instead of radiotherapy or chemotherapy, and felt the vitamin was a vital part of what was working. One person's story shouldn't be taken as important in terms of any health advice, and only a clinical trial can establish whether vitamin C can help with cancer. But what's perhaps more surprising is that even a clinical trial, as important as it is, doesn't always lead to a final and definitive answer; sometimes, the results of a clinical trial can still be argued about.

In 1979, Charles Moertel at the Mayo Clinic in Rochester, Minnesota, reported a well-designed full-scale clinical trial and found no benefit of vitamin C to cancer patients.[19] Nevertheless, Linus Pauling continued to argue the point. Patients in Moertel's clinical trial had been given relatively aggressive treatments beforehand, so Pauling suggested that their immune systems were probably too weak, or at least in a state that was harder to boost.[20] Also, the clinical trial had focused on whether or not vitamin C shrank the size of a person's tumour, whereas Cameron's initial observations in Glasgow had also considered how patients felt – did they seem to gain energy, for example? So it was possible that vitamin C has an effect not captured in the precise metrics that the clinical trial measured.

In a second trial, Moertel made sure to use patients who had not had any chemotherapy or radiotherapy beforehand. Once again the results showed that vitamin C offered no benefit.[21] But Pauling wouldn't back down, this time claiming that the problem was that the clinical trial had been set up to look for a decrease in a person's tumour within weeks. This was not appropriate, said Pauling, because vitamin C doesn't work by killing cancer cells directly, and so takes time to have an effect.

What this highlights is that clinical trials are vital, but also that they have limitations. A lot depends on the details: what's being measured, who the control group are, the timing and doses, any prior treatments the patients have had, and so on. Sometimes it's obvious that something has worked well and should be used widely as soon as possible. At other times it's still debatable even after a trial has concluded, as to what the results mean. Linus Pauling was driven by his own agenda, and fighting a very personal battle, but he was right to emphasise just how difficult it is to establish what really works.

This is true even for new medical treatments. In 2010, a specific type of immune therapy called checkpoint inhibition was shown to help some patients with metastatic melanoma, a type of skin cancer that has spread to other parts of the body.[22] But the treatment almost never got approved, because, even when working well, it does not always cause tumours to shrink straight away. Sometimes, indeed, they actually get bigger to begin with, on account of immune cells infiltrating the tumour. Only later, when the infiltrating immune cells have killed many cancer cells, does the tumour shrink. It was down to some observant clinicians who realised that even though checkpoint inhibitors looked like a failure at first, with tumours still getting bigger, in the long term some patients still benefitted. As a result, the criteria by which a cancer treatment was deemed successful were changed,[23] such as allowing an increased time for a treatment to work.[24]

Linus Pauling died in 1994. Late in life he claimed vitamin C could help combat HIV, stop the negative effects of ageing and help prevent heart disease. None of this has been proven, but he was listened to nonetheless and always widely reported in the media. People simply liked what he said, because for anyone afflicted with a difficult medical problem, or indeed just seeking better health, he offered something hopeful and simple. It is something to guard against, especially when we are desperate for something to work.

Is vitamin D good for our immune health?

Unlike vitamin C, vitamin D can be produced by the human body, in skin exposed to sunlight, as well as obtained in one's diet or taken as a supplement. Foods rich in vitamin D include oily fish, such as salmon, mackerel and sardines, egg yolks and fortified cereals or margarine. Also, unlike vitamin C, vitamin D can be stored in the body too, for about two months. And for vitamin D there is stronger evidence of an effect on immune health.

After being synthesised in skin, or having been eaten, vitamin D is converted to another form in the liver, which is then its most abundant form circulating in blood. In the kidneys, this is converted again to its most active form. But as well as being handled in the liver and kidneys, vitamin D can also be directly taken up and processed by immune cells. This means that there can be a high local concentration of vitamin D and its derivatives around immune cells, separate to the body's blood level. Many experiments have shown that, in a lab dish, vitamin D tends to dampen the activity of specific immune cells. At first this might sound like something you don't necessarily want to happen – a negative. But actually it is extremely important for an immune response to resolve itself over time, to quieten down when a threat has been removed.[25] If the immune system stayed in a heightened state even after a threat has been removed, there is more chance that there will be collateral damage to the body. In any case, you wouldn't want to stay in a feverous state for longer than needed. The consequences of vitamin D deficiency on immune health, though, are multi-faceted and complicated, as it has multiple effects on different kinds of immune cells, including their ability to move into different parts of the body.[26]

If we look at population-level analyses rather than lab dish experiments, low levels of vitamin D have been linked with an increased susceptibility to infections. In 2021, an analysis of forty-three randomised controlled trials involving nearly 50,000 people found that taking vitamin D supplements helped protect against respiratory infections, a finding likely to relate to vitamin D being important in the body's innate response to germs, including the process by which immune cells destroy bacteria.[27] But the effect is small: thirty-three people would need to take vitamin D supplements for a whole year to prevent one single infection.[28]

During the Covid-19 pandemic there was much discussion about whether vitamin D could help, as early on people suffering the most from Covid-19 were identified as almost twice as likely to be deficient in vitamin D.[29] When reporting this, *The Times* newspaper did clarify that a correlation is only a correlation, that all sorts of other traits and behaviours could be linked to people with low vitamin D levels, any of which could relate to problems with Covid-19. But 'given how little it costs', *The Times* also commented, 'what have you got to lose?'[30]

We now know that some symptoms of Covid-19 are caused by an over-zealous or over-long immune response to the infection. So if vitamin D helps calm the immune system, this could feasibly link low vitamin D with worse symptoms. On the other hand, and more importantly, a clinical trial involving 6,200 people did not find any evidence that vitamin D supplements helped with the duration or severity of Covid-19 symptoms.[31]

Much clearer is vitamin D's link with autoimmune disease.

Several lab experiments have indicated that vitamin D helps stop unwanted immune responses against healthy tissues, by dampening down immune activity. This fits with population studies, such as one which followed over 187,000 women for twenty-one years, of which 173 happened to develop multiple sclerosis. The study found that women using vitamin D supplements had a 40 per cent reduced risk of developing the disease.[32] On its own, this study following the fate of people isn't conclusive, because many other factors will also correlate with whether or not people are taking vitamin D supplements, such as age or smoking.[33] Hence the importance of a randomised trial published in 2022. In this five-year long trial involving over 25,000 participants of average age sixty-seven, vitamin D supplements reduced the risk of developing autoimmune disease, including rheumatoid arthritis and psoriasis, a skin condition where immune cells attack healthy skin cells, by 22 per cent.[34]

Still, there's something to bear in mind. A 22 per cent reduction in the risk of autoimmune disease sounds like a lot, and something we should act on immediately. But the actual numbers of people developing an autoimmune disease fell from about 12 in 1,000 to 9.5 in 1,000. This is the same result just said a different way, and it doesn't sound as dramatic. Per cent changes and numbers don't lie, but the way you tell it counts. Also, the average age of participants in this study was sixty-seven, and we don't know whether the effects of vitamin D would be magnified or diminished in a younger group.

As with vitamin C, it is possible that vitamin D has a small effect on average but could be more beneficial for certain individuals. Indeed, whenever skin cells are hit by sunlight, some of us are genetically predisposed to produce more vitamin D than others. It seems that genetic factors play a bigger role in determining our vitamin D levels in the winter than in the summer, when people's vitamin D levels are much higher anyway.[35] Those of us predicted to have naturally higher levels of vitamin D are less likely to get psoriasis.[36] Plenty of other evidence also links a deficiency in vitamin D with psoriasis, which fits with some people's psoriasis symptoms flaring up when less exposed to sunlight, and could result from vitamin D directly affecting the health of the upper layers of skin, or the immune system itself, because psoriasis arises from the body's immune system not being held back appropriately from attacking healthy skin cells.[37] Crucially, however, trials testing whether vitamin D supplements alleviate the condition show mixed results.[38]

This highlights something important: even if a particular nutrient deficiency is robustly linked with an illness, that doesn't guarantee that supplementing the nutrient will resolve the problem. The disease process may have gone on for too long to be easily reversed. Or there could be other confounding factors which need to be tackled too. Or an underlying issue behind the nutrient deficiency might be the important thing, not low levels of the nutrient itself.

Having said that, evidence does point to vitamin D being vital for immune health generally. What's more, some people will not make enough of this vitamin from sunlight alone, depending on various factors including where they live, how they live and what time of year it is. In the UK, for example, the guideline is that during the autumn and winter, sunshine is unlikely to provide all the vitamin D you need. As such, care should be taken to get some from food or a supplement.

What about other vitamins and nutrients?

Countless other nutritional factors have been linked with immune health. Vitamin A is well established as being important for the development and proper functioning of many different types of immune cell. It is also important in our physical defence, playing a role in the body's production of mucus. In adults, vitamin A deficiency has been linked to an increased risk of tuberculosis;[39] in children, vitamin A deficiency is a risk factor for severe symptoms with measles, pneumonia and diarrhoea.[40] The World Health Organization recommends giving vitamin A supplements, for two days, to children with measles if they live in an area where vitamin A deficiency is prevalent. A systematic review in 2005 showed that this significantly reduces the risk that children under two will die from measles.[41]

Omega 3 fatty acids, a type of healthy fat found in salmon, walnuts and eggs, for example, have also been linked with immune health and, at least in a lab dish, can directly dampen the activity of immune cells called macrophages (which are especially good at regulating the activity of other immune cells), and reduce inflammation.[42] More generally, several chronic inflammatory diseases, including type 2 diabetes, non-alcoholic fatty liver disease and cardiovascular disease, have become more frequent alongside increased consumption of processed foods, sugar snacks and drinks. A direct link is suggested by the fact that animals fed on a so-called Western diet show elevated markers of inflammation.[43] But our understanding of why is still vague. One hypothesis is that a so-called Western diet, with a high intake of processed foods, saturated fats and simple carbohydrates, provides relatively low levels of raw materials that bacteria in the gut need to thrive and produce the molecules which regulate immune activity (such as butyrate, which we will discuss in the next chapter).[44]

Currently in vogue is the idea that some foods and drink are themselves anti-inflammatory, such as various vegetables, including broccoli, peppers and tomatoes, and green tea. The spice turmeric is also widely considered as anti-inflammatory, on account of it containing the molecule curcumin. More broadly, a so-called Mediterranean diet, rich in grains, nuts and fresh produce and low in red meat, is also considered anti-inflammatory. The idea here is that foods rich in molecules with anti-inflammatory activity may be beneficial against diseases associated with inflammation. There is certainly something in this, as adherence to a Mediterranean diet correlates with an improvement in various health indicators, such as a 6 per cent decrease in a person's risk of dying from cancer.[45] But there are other issues here to bear in mind: for one thing, everyone's diet is very complex, making it extremely hard to test the impact of anyone changing their diet to include more or less of one or a few foods; and for another, a lot of evidence for the effect of dietary changes derives from markers of inflammation in blood samples, which don't clearly relate to actual clinical outcomes.

Alcohol consumption – not exactly a nutrient, alas – has also been studied a lot for its effect on immune health. Heavy alcohol use is certainly a risk factor for all sorts of diseases, including cardiovascular problems and cancer. But there is controversy over whether there are benefits from drinking moderate levels of alcohol. Studies that have found any hint of benefit tend to feature prominently in the media. Some studies have found that a little alcohol can help prevent development of rheumatoid arthritis[46] – but others have come to the opposite conclusion: that any level of alcohol consumption increases risk.[47] A meta-analysis of twenty-nine studies, involving over 25,000 people, found no evidence for any protective effect of alcohol in osteoarthritis[48] (the most common form of arthritis, involving wear of the cartilage that caps bones in your joints, while rheumatoid arthritis is a disease in which the immune system attacks the joints.) Even if a little alcohol can have some benefit, the level at which this becomes detrimental will almost certainly vary from person to person, making any uniform guideline extremely problematic.

What can we conclude?

There are several messages from what we've covered here. Vitamin C won't really help much with colds, and there is good evidence that vitamin D is important in immune health generally. But also, from all the twists and turns in the decades-long debate about vitamin C and colds, as well as what we know about some other nutrients, a lesson emerges: how hard it is to really know. At one level this is unsatisfying. But at another, this is the greater truth. Which brings us back to another long-held truth: don't latch on to any one thing – a healthy diet has to be a *varied* diet.

There are many grandiose questions for science to still solve: the origins of life, the endpoint of the universe, how memories are stored in the brain. Something as down-to-earth as nutrients and immune health and who needs what when, is not usually talked about in the same way. Somehow it is not considered as elaborate or as monumental. But it is just as knotty. We have some indications and some snippets of answers, as we do for these other weighty issues, but the best methods for tackling nutrition and immune health have yet to be developed.

In 2023, most research institutes and centres funded by the US government continued with a budget similar to the preceding year's. For one office, though, President Biden doubled the budget: the NIH Office of Nutrition Research. He knew, or had been advised, that nutrition science is ripe for transformation. Or in desperate need of it.

2

The Aliens Living Inside You – or, does the microbiome affect immune health?

Over the years, I've been asked many times if I would consider being involved in promoting various things, including stores selling food supplements, a high-end gym chain and even a dating agency (my first book included a section on whether the immune system affects who we find attractive). Most recently, I was asked if I would consider promoting yoghurts containing live bacteria which are advertised as supporting immune health. I have never said yes, but every time I'm asked, the question of scientists promoting things plays on my mind.

It's a tricky issue: on the one hand, scientists could explain how, when or why a product is helpful, but on the other hand, any money involved could seed bias. Gifts from companies to medical professionals are carefully controlled, because they might influence treatment practices for patients, even unwittingly. All scientific papers and talks must disclose who funded the research. For me personally, there's another problem: I would need to dig into the subject to know what I really think about whatever product I was being asked to promote. And because there's so much depth to immune health – I've spent thirty years studying immunology and still don't know the half of it – it seems unlikely that I could ever get behind a simple slogan.

Yet simple slogans are rife. Adverts and packaging for

food and drinks commonly feature scientific-sounding phrases like 'supports immunity', 'protects against colds' and so on. It's often not clear who's saying it. In one case, on the packaging of one probiotic yoghurt, the statement appears to come from fictional yellow creatures called 'the minions'. There are more cunning promotions too, permeating our culture. In 2023, for example, a popular computer game called Fortnite included a feature whereby characters can drink a virtual probiotic yoghurt drink to boost their resilience.[1] Having trouble hunting zombies? Get your character to an 'Immunity Station' to pick up a yoghurt drink, just like you can in real life.[2] But of course, to really understand whether you may benefit, or who may benefit, we need to know all sorts of details: where the idea comes from, the evidence, the caveats, the nuances, the current frontier of knowledge, and what looks feasible in the near and distant future. Which is why I am writing this book, not working with the minions. Leave aside the cartoons and gameplay, gimme some truth.

To get to grips with all this we first need to understand how we are affected by the microbes that live inside us; then we can ask whether these microbes can be changed or targeted in some way to improve our immune health, with a probiotic yoghurt or anything else. This is no easy task, because there are trillions of microorganisms – bacteria, fungi and viruses – that live in and on the human body. They're found in all sorts of places; in the gut, mouth, eyes, nose and all over our skin. Where to begin?

Well, historically at least, the idea that the body's resident microbes might affect immune health has its roots in another big idea: that growing up in a somewhat 'dirty' environment, such as a farm, may help immune health, and especially for preventing the development of allergies caused by an unwanted immune reaction to something harmless like pollen or dust mites. So let's begin here: does growing up on a farm help immune health, and if so, why?

Is a little bit of dirt good for us?

One week in March 1958, over 17,000 babies born in the UK were enrolled in a long-term scientific study. When they had reached the age of eleven, their parents were asked if their children had experienced allergies in the last twelve months. Twelve years later, the babies, now adults, were asked about their own allergies. David Strachan, based at St George's Hospital in London, studied the results to understand why people develop allergies.

He found a link between family size and whether a person developed hay fever. The bigger the family, the less likely they would develop the allergy. Having older siblings, especially, decreased the likelihood. Knowing that larger households generally experience more infections, Strachan reasoned that increased infections in early childhood might protect against allergies. This went against the prevailing view of the time, which was that an infection might cause an allergy, not protect against it.

Strachan himself has always emphasised that this idea did not come from him alone; he was influenced by earlier work from David Barker at Southampton General Hospital, who was trying to understand why rates of acute appendicitis were increasing in the UK.[3] One idea was that a lack of dietary fibre from cereals, vegetables or fruit, was causing increased appendicitis – the so-called fibre hypothesis. To test this,

Barker compared the frequency of appendicitis in different parts of the UK against the types of food consumed in that area, such as the total average weight of vegetables people ate, and so on.[4] He found no evidence for the fibre hypothesis. Something else had to be happening.

Appendicitis is usually a sudden problem, triggered by an inflammation that causes the appendix to swell, sometimes even to burst. In other words, an over-active immune reaction is often the problem. So, Barker wondered what might be increasing the frequency of over-reactive immune responses, and if fewer encounters with bacteria when we're young, caused by modern practices of hygiene, could increase the risk of us over-reacting to bacteria later in life? If that were true, he reasoned, then increasing sanitation and hygiene might account for the increasing risk of acute appendicitis.

In 1989, building on Barker's ideas, Strachan published his findings about allergies in the *British Medical Journal*. He named the idea that the immune system needs an exposure to germs to develop properly, the hygiene hypothesis.[5] Nowadays, the hygiene hypothesis has come to denote much more than something relating to just appendicitis or allergy. It is invoked for all sorts of issues, such as the rising frequency of autoimmune disease and even our susceptibility to viral infections.[6]

There are many related issues too, such as the use of antibiotics in children, or by mothers during pregnancy.[7] Children and pregnant women must use antibiotics when needed, because they save lives against harmful bacterial infections, but antibiotics don't only kill the harmful bacteria they were prescribed for, they also kill some of the body's normal resident bacteria. So if exposure to non-harmful bacteria helps a baby's immune system develop in the womb, then this might be affected by antibiotics taken by the mother. There has been some suggestion that antibiotic use is linked with childhood asthma, though we need more research here to understand what's really going on.

Really, the 'hygiene hypothesis' has turned out to be an unfortunate name, because it's not cleanliness that's the problem – not in the everyday sense of the word.[8] Modern standards of sanitation are certainly not bad for health overall; in fact the complete opposite. The WHO estimates that 1.4 million deaths occur every year because of inadequate drinking water, sanitation and hygiene,[9] making hygiene one of the most important lifesavers of all.[10]

It is also not true that personal hygiene is a problem. There is no evidence, for example, that showering or bathing changes our overall risk of allergies or autoimmunity; nor does the use of soap, detergents and cleaning products correlate with the prevalence of asthma, hay fever or eczema.[11] Again, the opposite is the case: cleaning to get rid of household allergens, such as those produced by dust mites or mould, helps with allergies. So hygiene is generally good for our immune health (as it is for health more broadly). What Strachan really meant back then – and we know has some truth to it – is that there is something beneficial about a 'dirty' environment exposing us to some microbes. It's worth dwelling on this for a moment to appreciate how radical an idea this was, and still is. Even today we tend to think of microbes as harmful and capable of causing disease. But Strachan was suggesting the opposite: that some microbes help protect us from disease.

So is dirt good for immune health? Well, the fact that children who grow up on traditional farms are less likely to have asthma, hay fever and other allergies certainly fits with the hygiene hypothesis.[12] And this is not a small effect: it amounts to something like a 25 per cent reduction in the likelihood of allergic asthma developing.[13] To understand why, two farming communities in the US have been closely studied. The Amish and the Hutterites have similar ancestry, similar diets, large families and get childhood vaccinations. Yet Hutterite children are around four times more likely than Amish children to develop asthma. An important difference is that Amish farms use traditional methods on single-family dairy farms, while Hutterites use communal mechanised farming. This means that Amish children live much closer to animals and the environment where animals live.[14] An analysis of immune cells in the blood of both groups of children found that certain types of immune cell were continuously stimulated at a low level in Amish children, more so than Hutterites. This fits with the idea that the immune system of Amish children is being continuously tickled by the presence of bacteria in their environment. In other words, Amish children live potentially somewhat less 'hygienically' – are more exposed to germs – and get asthma less frequently.

What's more, and quite astonishingly, dust samples from Amish homes, but not Hutterite homes, suppress the symptoms of asthma in mice.[15] An analysis of the molecules present showed that Amish farm dust contains a number of different molecules from animals, plants, bacteria and fungi.[16] It could be that some of these molecules, perhaps in a particular combination, have a direct effect on immune health. One thesis is that an exposure to germs may be important in training the immune system, effectively setting a higher threshold for it to switch on, making it is less likely to respond to harmless things like pollen or dust mites. But this is speculative – we don't know for sure. A related possibility is that mud or thermal water baths *could feasibly* impact immune health, perhaps by exposing us to beneficial harmless microbes, affecting the microbial communities on our skin, perhaps, but there is no proof, and other effects would come into play too, such as baths helping reduce stress (which we will come to later).

In fact, the protective effect of farms *may* have little to do with the hygiene hypothesis, because of other factors. For example, drinking unprocessed, raw cow's milk also correlates with a lower risk of asthma or hay fever developing.[17] However, raw milk can also contain disease-causing bacteria, so a systematic trial with young children has not been done, for safety. As to what might be special about raw and unprocessed cow's milk, we can only guess.[18] In general, milk contains an abundance of bio-active molecules: vitamins A and D, various protein, fat and sugar molecules, and any number of components we scarcely understand. Human breast milk contains small spheres made up of fat molecules surrounding a core of thousands of different types of protein molecule, and we know very little about what they do.[19]

A further complication is that careful scrutiny of the levels of asthma on farms reveals something of a paradox. While growing up on a traditional farm protects against asthma, working on one has the opposite effect, and increases the risk. How can this be? Essentially it comes down to asthma not being one thing. Asthma is an inflammation of the airways that leads to coughing or wheezing, chest tightness or shortness of breath, and a variety of underlying processes can cause this. There are two broad categories: allergic asthma, caused by an immune reaction to things like pollen or dust mites, and non-allergic asthma, caused by something else, such as air pollutants, dampness and mould or vigorous exercise. On traditional livestock farms there is less risk of *allergic* asthma, but an increased risk of *non-allergic* asthma.[20] Adult farm workers are routinely exposed to dust and microbes, gases and other irritants, sometimes in relatively confined farm buildings, which increases the risk of respiratory problems generally, including non-allergic asthma.[21]

So it is *not* a myth that farm life offers something important for immune health — but unfortunately it *is* a myth that we can act on this now, for example to tackle allergies. We can't deliberately expose young healthy children to farm dust or raw cow's milk, because each constitutes any number of possible side-effects and dangers, and the level of protection they would give is unclear.

Instead, we need to work out precisely what's going on. Which specific molecules are important, where do they come from, how exactly do they affect us, do they affect all of us, or just some of us, and so on? Only then can we develop new supplements, medicines or guidance. In the meantime, a recent and crucial shift in our thinking about the hygiene hypothesis, and its relationship to immune health, has brought the realisation that something big was missing at the outset, something David Strachan could not possibly have foreseen: the importance of gut microbes.

Happy gut, happy immune system?

For all the microbes on and inside your body, the ones living in your gut — the gut microbiome, sometimes just called *the* microbiome — have been studied the most, and are nowadays at the forefront of public awareness. As it happens, the stomach itself is relatively low in microbes, because the acid there kills many and stops others from multiplying. Just below the stomach, many kinds of microbes live in the small intestine, but they thrive best of all in the 1.5 metres of gut beneath that: the large intestine, the final stretch of food's journey.

Analysing what's inside a person's gut directly is hard. So whenever you read that someone's gut microbiome has been analysed, what's really happened is that – as delightful as it sounds – their faeces have been analysed for the bacteria they contain. (Faeces contain many kinds of microbes, including bacteria, yeasts and viruses, but it's the bacteria which have been studied the most.) In fact, 25–50 per cent of faeces are made up of bacteria. From such analysis we know that gut bacteria vary a lot from person to person, change during puberty and pregnancy, and are affected by factors like our diet and where we live. So to test whether the microbiome affects immune health, the most common experiment has been to compare the contents of people's faeces with whether they are suffering from something related to immune health, such as an autoimmune disease. One thing immediately evident is that, with so many things affecting the make-up of a person's microbiome, it is certainly not all down to diet, let alone one small part of anyone's diet, such as whether they drink a probiotic yoghurt every so often.

One study compared the faeces of children in Finland, Estonia and Russia, because autoimmune disease is more common in Finland and Estonia than in Russia. Children had their faeces analysed every month for three years, while parents filled out questionnaires about diet, allergies, infections and drug use.[22] This huge undertaking showed that where children lived was a major factor in the types of bacteria that made up their gut microbiome. Independent of anything identifiable about their diet, whether they had recently used antibiotics or anything else, certain types of bacteria were especially high in Finnish and Estonian children, while others were more common in Russians.

It turns out that a molecule from the outer coating of a type of bacteria common in Russian children is well known to be potent at switching on immune responses. The same type of molecule, but from the outer coating of bacteria more common in the gut of Finnish and Estonian children, actually tends to switch off immune cells. So, conceivably, bacteria commonly found in Russian children could help protect them against autoimmunity – an unwanted immune response against the body's normal healthy cells and tissues – by switching on a low-level of immune response early in life, which helps train the system to respond appropriately later in life.[23]

While the microbiome may affect the likelihood of autoimmune disease developing, does it affect immune health in other ways too? The answer is yes, and another example comes from cancer patients. One type of immune therapy, called a checkpoint inhibitor, works by taking the brakes off an immune response, helping some patients' immune system fight their cancer. We will discuss how this therapy works in detail in the next chapter, in the context of obesity. Important here is that checkpoint inhibitors only help with some types of cancer, including lung, liver, kidney and skin cancer, and even then they don't work for everyone. Surprisingly, one thing which correlates with the likelihood of a patient responding well is the composition of their microbiome. But there's a catch. Different studies have arrived at different conclusions over what the beneficial component is.

In treating epithelial cancer, a type of cancer that can affect your skin, breasts, kidney, liver, lungs or pancreas, for example, the presence of certain bacteria correlated with the success of checkpoint inhibitors.[24] But a study of melanoma patients found other bacteria to be important, which makes it extremely hard to understand what's happening.[25] Evidently, no single species of bacteria is the magic ingredient.[26] For now, then, there isn't any simple test that can be done on a person's faeces to know whether they are likely to benefit. Nevertheless, it is certainly possible that in the future, manipulating the microbiome, or using specific molecules normally produced by the microbiome, in combination with checkpoint inhibitors or other immune therapies, may become a medically important strategy for treating cancer.[27] This is a hot topic, with several clinical trials and studies ongoing.[28]

All in all, there's scarcely any state of human health or disease that hasn't been linked with the gut microbiome.[29] Even neurological conditions such as Huntington's, schizophrenia, depression and autism.[30] But it is very difficult to work out whether variation in a person's microbiome *directly* causes any particular condition, worsens the problem or is something that just happens.[31]

One way of testing this is to use mice. You'll have your own views in terms of the ethics and morality, but the study of the microbiome and immune health has benefitted from experiments with animals because there is less complexity compared with correlations across people. One famous and important experiment used mice with a genetic mutation so that their T cells couldn't develop normally, which made them susceptible to an inflamed gut condition called colitis. Surprisingly, when a second group of healthy mice were put in the same cage as the mutant mice, they got colitis too.[32] We know that mice living in the same cage tend to end up having similar gut microbes, so the simplest explanation is that the microbiome caused this gut inflammation. It doesn't mean that human colitis is inevitably the outcome of a change in a person's gut microbiome, but it does open up the possibility of helping colitis patients with something that manipulates their gut microbiome. We're not there yet, but the research continues.[33]

To test precisely for the effect of the microbiome on immune health, mice kept in a sterile environment all their life have been fed food that has been irradiated, so they have never been exposed to any microbes. Now these mice can be deliberately exposed to them, including those derived from humans. One experiment gave sets of mice microbes extracted from the faeces of identical twins, one of whom had multiple sclerosis and the other not. Even though the mice used were already predisposed to developing a disease like multiple sclerosis, it's striking that those receiving bacteria from the faeces of the twin with multiple sclerosis were much more likely to develop symptoms themselves.[34] Similarly, microbes isolated from the faeces of people with inflammatory bowel disease cause symptoms to be worse in susceptible mice.[35]

Could something as complicated as autism even be affected by the gut microbiome? Of course, studying something as complex as autism in animals is fraught with problems, as it's extremely hard to know which, if any, criteria are applicable. Nevertheless, a single gene alteration in mice causes them to show behaviours reminiscent of autism, including less social interaction with other mice. Microbes isolated from the faeces of these mice could then be transplanted into other mice. (The receiving mice had previously been treated with antibiotics to reduce their own normal gut bacteria.)

Amazingly, this transfer of microbes caused autism-like behaviours. What's more, the converse effect was seen too; when mice with autism-like behaviours were given microbes from other mice, their tendency to avoid social contact became less pronounced.[36] These results, and many others like them, are good evidence that the microbiome has the potential to directly affect all sorts of conditions or alleviate the severity of symptoms.

It's perhaps not so surprising that something like colitis would be affected by gut microbes, because the disease itself manifests in the gut. But as we've seen, the gut microbiome also affects diseases and conditions like multiple sclerosis that have nothing to do with the gut or digestion, the outcome of cancer therapies and behaviours associated with autism. How?

One way is by gut microbes extracting and producing nutrients which affect the whole body. Going back to the mice susceptible to autism-like behaviours, it turns out that their microbiome produced lower than normal levels of vitamin B_6. To test whether this might underlie symptoms, mice not otherwise susceptible to autism were deliberately deprived of vitamin B_6 for two weeks, which did indeed lead to decreased social interaction. But the mice could have become less social because they were tired; or perhaps a change in their sense of smell would affect their ability to recognise one another; or they might have become hyperactive, or more impatient . . . So it doesn't prove that low levels of vitamin B_6 directly relate to autistic-like behaviours. And just as importantly, *supplements* of vitamin B_6 had no effect on mice genetically susceptible to autistic-like behaviours.[37]

As we've already said, nutrition science is hard. I know we want a simple takeaway — imagine how many books I could sell if I declared vitamin B_6 deficiency to be linked to autism, by cherry-picking some of the experiments performed with mice while ignoring other data and all the caveats! The truth is much less of a soundbite: the gut microbiome can affect the whole body, and one way is through affecting the production and extraction of nutrients.

Whole-body health is also influenced by something much more subtle: gut microbes' complicated interactions with one another. One type of gut bacteria might thrive on the waste products of another; another might produce molecules poisonous to other types. If two species rely on a common resource to live, a competition arises, and some types of bacteria are so fiercely competitive that they actively kill others.[38] Microbes are continuously multiplying and dying, but eventually, hopefully, a relatively stable level is often maintained.[39] That's not to say change doesn't happen, but in general a person's microbiome is going to look more like the same microbiome three months later than like someone else's.[40] This is important for our health, because stability in the microbial community lessens the chance of one disease-causing species running riot.

Indeed, loss of that stability means trouble. *Clostridium difficile*, or *C. diff*, for example, can cause a variety of problems, ranging from diarrhoea to a high fever, and occasionally it can be fatal. A person infected with it can suffer these symptoms, as you might expect – yet surprisingly, this type of bacteria is always present in some people's gut without them being ill or even knowing it's there. In such cases its numbers are kept under control by the overall ecology of the microbiome, and by the immune system too. But any change in a person's microbiome has the potential to give *C. diff.* an opportunity to flourish. Occasionally this can happen when taking antibiotics, which, as we've already seen, destroy the dangerous type of bacteria they were prescribed for, but also have off-target effects on others. Sometimes the overall ecology of the gut microbiome is affected, allowing *C. Diff.* to explode in number.[41]

Evidently, microbiome stability is important. But it takes time to develop. Initially, the make-up of our microbiome is influenced by the nature and timing of our birth and mother's diet. Soon after birth, it starts being affected by what we're fed and the environment we live in.[42] For example, antibodies from breast milk directly influence the composition of the gut microbiome.[43] As we grow up, different kinds of bacteria increase and decrease in waves until an adult-like ecology develops.[44] Alas, we do not yet understand these processes enough to give advice to an expectant mother, or about breast feeding, or what to eat when we're young.

Until recently, it was thought the microbiome developed most in the first three years of life, and then settled into a more adult-like structure. But new research suggests that it continues to change considerably throughout our youth and teenage years, evolving more slowly than we thought. The old view was perhaps just down to more studies being carried out on young children than older children and teenagers.[45] But even if the precise age at which our microbiome settles is a little unclear, what's important is that its overarching structure – our own levels of the thousand or so different bacterial species living in the human gut – *is* influenced by what happens to us when we are young. This also fits with the hygiene hypothesis: that the environment we grow up in has a lasting impact on immune health.[46]

How does the makeup of our microbiome affect our immune system?

Everyone's microbiome is in constant dialogue with their immune system. Immune cells surround the intestine, and one job they have is to be poised to directly attack any microbes that try to leave the gut and invade the body. Although outside the gut, many of these immune cells have small extensions which squeeze in between the cells making up the 'skin' of the intestinal tube itself, to contact gut microbes on the inside. These immune cells produce antimicrobial molecules which accumulate in a protective mucus lining.[47] Other cells not normally considered immune cells as such, including those which make up the 'skin' of the intestine tube, have their own ability to recognise gut microbes, and the secretions they produce can recruit immune cells to the intestine as and when needed.

The intestine itself, at over 9 metres long, is not just one long, unvarying tube.[48] There are all sorts of distinct neighbourhoods and communities, crypts and cavities inside, each with their own set of immune cells, other cells, and microbes.[49] All along, the immune system must refrain from reacting against the good-for-us microbes while being ready to attack any which pose a threat. So, how does the immune system really know what to react against and what to leave alone?

It's not that immune cells have some inbuilt instruction to know that one type of bacteria is dangerous, and another not. Instead, whether immune cells decide to respond depends on various things, including where in the body the bacteria are, if there are other signs of trouble, or if that type of germ has caused a problem in the body before. For example, the exact same type of *E. Coli.* bacteria that live happily in our gut will cause disease if they get into the urinary tract, and then an immune response is needed.[50]

To really understand how the microbiome affects our immune health, we need to focus on specific immune cells called regulatory T cells or T regs (said like the dinosaur T. Rex). These are immune cells which specialise in turning *off* other immune cells. In other words, T regs have the task of ensuring that an immune response doesn't happen against something which does not warrant it, or that an ongoing immune reaction doesn't overshoot. A molecule called butyrate is instrumental here.[51] Butyrate is produced during the chemical process by which many gut bacteria gain energy,[52] which matters because butyrate directly affects the activity of T regs.[53]

Many types of normal gut microbe produce butyrate to help T regs keep the immune system from attacking them. But T regs also work throughout the body, and help prevent allergies and autoimmune diseases as well. So the production of butyrate (and other molecules like it) is one way in which the gut microbiome is directly linked to our susceptibility to allergies and other illnesses that come from an unwanted immune response.

Again, some of the evidence for this comes from research with animals. Mice given a high-fibre diet increased the overall numbers of certain types of bacteria in their gut, which in turn produced high levels of butyrate and correlated with them being less likely to develop asthma.[54] Similarly, microbe-derived butyrate lowered the risk of autoimmune arthritis developing in mice.[55] More importantly, there is evidence for this in humans. An analysis of the microbiome in people newly diagnosed with rheumatoid arthritis showed they had low levels of the type of bacteria that produce high amounts of butyrate. This fits with the theory that butyrate, and the activity of T regs, help keep the immune system from doing something it shouldn't. Similarly, children with allergies tend to have microbiomes with low levels of bacteria which produce butyrate.[56] Young children with allergies also tend to have low levels of butyrate in their faeces.[57]

How exactly do T regs help our immune health, then? T regs have at least four modes of action. First, they secrete protein molecules which directly suppress the activity of other immune cells. Secondly, they can disrupt nearby immune cells by using up a resource they need. Thirdly, they connect with a specific type of immune cell called a dendritic cell, that's especially potent at beginning an immune reaction, giving these cells a specific signal to quieten down.[58] Finally – and dramatically – T regs can kill off other immune cells.

Given such power, it's perhaps not surprising that T regs have caught the attention of the pharmaceutical industry as well as academics. Future medicines could harness or unleash the power of T regs to stop unwanted immune responses in autoimmunity or allergies, and in many other situations. Sepsis happens when the immune system overreacts to an infection, leading to unwanted tissue damage, organ failure and sometimes death. Perhaps surprisingly, many neurological problems also involve unwanted immune responses, including stroke, spinal cord injury, traumatic brain injury, multiple sclerosis, Alzheimer's and Parkinson's. While there's still a lot we need to understand about T regs, it's not implausible to suggest that in future they could be harnessed in treatments to help stop or delay deterioration in such conditions.[59]

Something else is important here too. The skin, lungs and mouth have their own microbiomes too. We know far less about these communities, but it seems very likely that, for example, the microbes which occupy the lung will impact immune cell activity there.[60] There's also the fact that the microbes which live in our lungs are probably not entirely disconnected from those in the gut, or anywhere else in the body. We've seen how microbes interact with one another when they are nearby in the gut, but maybe the same process stretches out across the body, creating an entire corporeal microbial ecosystem. We do know that the make-up of both the lung and gut microbiomes correlates with the likelihood of various lung conditions, including asthma, and our susceptibility to lung infections.[61]

Overall, then, there is no question that the microbiome is important to immune health. So can we manipulate it to help, by the food and drink we consume or in any other way?

Does that yoghurt drink really do what it says on the bottle?

There are broadly two ways in which live bacteria in so-called probiotics could feasibly support immune health. The ingested bacteria could act directly on our immune system, or they could affect the microbiome, which could in turn affect the immune system. Several lines of evidence say both can happen. First, in lab dish experiments, many of the protein, fat and sugar molecules present in probiotic cultures trigger a reaction in a variety of immune cells.[62] There's evidence this can happen in the body too. For example, volunteers fed a particular strain of bacteria called *Bifidobacterium infantis* 35624 – a billion of them once per day for eight weeks – gained increased levels of T regs in their blood.[63] This shows a clear effect, but stops short of proving a medical benefit, because there's no easy way of telling whether people are now better protected against developing an autoimmune disease or allergy as a result of having higher numbers of T regs.

It is also clear that ingesting bacteria affects the microbiome. In one study, probiotic yoghurt consumed by patients with inflammatory bowel disease changed the make-up of their microbiome.[64] In another, the effect of antibiotics on the diversity of a person's gut microbiome could be circumvented if they took probiotics at the same time.[65] One large study of babies born prematurely showed their microbiome was affected by specific probiotic products.[66] Particular strains from probiotics persisted for months.[67] Interestingly, once a hospital used a probiotic for infants, bacteria from it were often detected in the faeces of nearby babies who had not yet been given it themselves. In other words, cross-contamination can happen through a shared environment, just as those mice housed together tended to end up with similar bacteria in their microbiome too, or people living in the same place share some features of their microbiome.

Importantly, though, in every human microbiome study there is a lot of variation between people. Changes can be detected on average, but some people are always affected more than others. Don't forget that the actual configuration of our own microbiome reflects so many things: all the microbes we have been exposed to, how they interact with one another, which types are tolerated by our own immune system, and if our intestine itself accommodates their presence. With a thousand or so different bacterial species able to live in the human gut, almost any number of stable microbiome communities are possible. With probiotics potentially causing lifelong consequences, therefore, much more needs to be done to understand individual variation[68] – something to bear in mind when you come across simplistic claims.

With good evidence that probiotics can impact a person's microbiome it'd be easy to assume this is important and positive and you should start taking them. But here's the thing: probiotics are pitched as a prophylactic, a preventative, against disease developing, which is difficult to establish, especially in people who are essentially healthy. If the numbers of T regs in a person's blood increase, for example, which may well be put down to taking a probiotic, it's hard to know if this would be enough to help with an allergy or other illness. And in helping or hindering our fight with an infection this could easily go either way. Some symptoms of infection come about from the immune system itself overreacting, in which case dampening immune activity may help, but of course we don't want the immune system entirely switched off if dangerous germs need to be fought. Achieving this balance is delicate, and it's still not obvious when probiotics help. Probiotics certainly can change a person's microbiome, but short of deliberately giving someone an infection, which is not ethically appropriate, it's hard to track or calibrate an improvement in immune health.

What might really help?

Vegetable fibre or supplements which encourage gut bacteria to multiply – so-called prebiotics – could also feasibly nudge the state of our immune system to our benefit, but it's difficult when nurturing one kind of bacteria to ensure that a closely related but detrimental species of bacteria doesn't also thrive.[69] A high-fibre diet is generally thought to encourage a balance of different kinds of bacteria to prosper in the microbiome – which would be healthy. On the flip side, we wouldn't want to ingest something that helps a dangerous kind of bacteria prosper in the gut. The role played by diet in countering a *C. diff.* infection, for example, remains debatable.[70]

Better understanding is needed before more technically complex products can be developed that are, hopefully, more proven to work. One way to make probiotics more sophisticated is to use genetically modified live bacteria. This is relatively easy to do, as we saw back in 1978, when bacteria were engineered to produce pure insulin for use as medicine. In that case, the bacteria were only being used in the industrial process of producing it, but essentially creating genetically modified live bacteria which could be added to food draws on the same technology.

Is anything proven to work with the microbiome?

There is one way of manipulating the microbiome which *has* been clinically proven to work against a specific illness. A faecal transplant. Yes, you read that correctly. Astonishing as it sounds, a fresh stool sample is collected from a healthy donor, whizzed up, sieved and delivered to a recipient via a long tube into their rectum. The faeces transferred can be from anyone – a partner, family member or a stranger (does the notion feel nicer coming from a partner rather than a stranger, or not?). This procedure really can work in the treatment of a *C. diff.* infection which is causing particularly extreme symptoms and might otherwise be fatal. One reason it works more easily than say, a food supplement is that direct insertion gets the bacteria where they are needed, and the full complexity of a healthy person's microbiome is being used. It may be a less palatable concept than consuming an 'immune-boosting' yoghurt every so often, but does at least show that a change in microbiome *can* be clinically effective.[71]

So what's the conclusion, and what do I do?

I refrain from participating in promotions or advertising because I think it takes a book, not a slogan, to present the appropriate detail. And I want you to see how fragmentary the evidence remains as to exactly *how* the microbiome might be used to help immune health, even if what is building points to genuinely exciting possibilities on the horizon.

But when I give talks, I do get asked: what do *you* do, in your own life? The truth is, things aren't straightforward here either. Intellectually, academically, I know that orange juice will not help me recover quickly from a cold, and know that a yoghurt drink claiming it 'supports your immune system' has not been proven to do so. But it doesn't mean I won't use them. It's as though there's a secret set of scales influencing my actions, with upbringing, culture and advertising weighing down one side, and all the scientific data I've come across on the other.

And it's not just me. A scientist whose research focuses on the gut microbiome once told me that when she had to take antibiotics for an extended period, she also adopted a diet especially high in fibre. Not because it's proven to help: just that it *might* help keep her gut microbiome stable by allowing normal gut bacteria prosper, and was almost certainly not harmful. The point is not that this is right or wrong, but that no one really knows. David Strachan has said his greatest frustration has been that in decades 'of epidemiological and immunological investigation . . . so little progress has been made' and that the idea of microbiome restoration or change has yet to be proven to work.[72]

Three takeaways, then. First, beware of the hype: no bacteria-enriched yoghurt drink has been proven in any large-scale clinical trial of healthy people to boost or support immune health. Secondly, experts don't really know either. Each of us is fumbling through the complexity in our own way. Thirdly, a lot of evidence shows that gut microbes are important for immune health – but only in the future will we know how to harness this.

3

The Evidence of Weight – or, does weight affect immune health?

Weight is a vexed topic. It's hard to remember how infinitely elaborate every single one of us is when every day we're bombarded with images and messages that makes us dwell on our bodies' crude shape. Fat-free, sugar-free, light, diet, zero, low-calorie, any number of clubs, online chats and apps, and big industries from fashion to dating: all link slenderness and weight loss to aspirational goals, a sense of achievement, ambition, self-control, even morality. Being a certain weight – thin but not too thin – has become a measure of success far beyond what it takes to be healthy. So before we discuss weight in the context of immune health, two general points I feel are important to make.

First, how much a person weighs – whether they're over- or underweight – is not really that person's fault. The food processing industry has perfected combinations of sugar and fat to be almost magically delicious and addictive, which in effect hack our taste buds and neural wiring. Meanwhile other companies, or just different branches of the same ones, are promoting an obsession with thinness in order to sell their diet drinks and foods. We are all affected. How could we not be? At the same time our body is being shaped by all sorts of other things too, in ways we're only beginning to understand: our socioeconomic and cultural background, levels of exercise, the eating habits we grew up with, how well we sleep, the composition of our gut microbiome, our genes and so on. What we do know is that many factors affecting our weight are either beyond our control or extremely hard to control.

The second thing is that, as the science of the human body advances and we learn more about its composition, we must beware of letting any metrics about it become more powerful than they should be. All numbers about the body conceal the person and their circumstances. Even something as simple as the body-mass index can be skewed by very high muscle mass. Also, all of us distribute fat in different places, which in turn influences our health. Of course, deciphering what's a supposedly 'normal' level can be helpful if it's perceived as something to act on, but labels can themselves trigger problems: disconnected from their actual biological meaning they become just something to worry about. Unfair generalisations can damage our sense of self-worth.[1] I'm not saying that we should abandon scrutiny of the human body – most people understand the benefit of using metrics about our bodies to arrest disease before it takes hold. But we need to be mindful of the impact body metrics or labels can have.

Being either overweight or underweight does affect immune health, but in surprising and significant ways. For example, it is not true that being overweight worsens our susceptibility to every disease, occasionally it could even be beneficial.

Does fat harm the immune system?

Once considered an issue confined to high-income countries, and perhaps specific cultures, being overweight is on the rise globally, including in low- and middle-income countries.[2] In fact, over a third of all adults – more than 1.9 billion people across the world – are considered overweight or obese.[3] It's not just adults. In 2019, an estimated 38.2 million children under five met the criteria for being overweight or obese. Indeed, most people today live in a country where being overweight or obese causes more deaths than being underweight. Not one country has ever managed to halt, let alone reverse, an increasing average body weight.

Being overweight or obese is linked to all sorts of diseases, including heart disease, musculoskeletal problems and many types of cancer. There are many reasons, a lot of which aren't specifically to do with immune health. For example, the risk of heart problems is related to an increased blood volume and cardiac output.[4] A build-up of fatty material in blood vessels increases the risk of stroke. Fat compressing the upper airways can contribute to disrupted sleep and snoring. However, some diseases, including type 2 diabetes, multiple sclerosis and rheumatoid arthritis, involve the immune system more directly, which begs the question: how does excess fat affect the immune system?

Body fat is made up of specialist cells below the skin which store fat for later when more calories are needed – fat being twice as energy-rich, gram for gram, as sugar. But fat cells do much more than store fat: they also produce hormones and other molecules which regulate other processes in the body. This is an important general point for all human biology: any tissue we associate with one particular function – muscles, say, which control movement, or fat, which stores energy – does a great deal more, because of all the molecules it produces, which then act on other parts of the body. Every part of the body multi-tasks, everything connected and interdependent.

An increase in body fat happens in two ways: fat cells can increase in number, and individual fat cells can get bigger. Where in the body fat cells tend to expand in number or size varies from one person to the next, and why is still basically a mystery. But wherever there's increased fat – either more cells, bigger cells or both – it'll be producing higher levels of molecules.[5] It's these molecules that directly affect our immune system, throughout the body.

Perhaps surprisingly, there is also a vast array of immune cells that live within body fat. If we have an excess of fat, the immune cells which reside within our fat tend to become more active. They also produce molecules, some of which drift out into the body. Various molecules produced by fat itself, and immune cells living within fat, can trigger unwanted inflammation elsewhere in the body, which in turn, increases our risk of disease such as cardiovascular problems or rheumatoid arthritis. One disease that has been studied a lot in the context of weight, as you probably already know, is type 2 diabetes.

Is it true that being overweight makes a person more likely to develop diabetes?

Globally, around 450 million people are affected by type 2 diabetes, and it is the ninth most common cause of death.[6] Symptoms can include feeling particularly thirsty or tired, urinating a lot, wounds taking longer to heal, and blurred vision. Many people have the condition without realising it. There is no cure (yet).

In diabetes, insulin is the key player. It is a hormone produced in the pancreas which helps maintain normal blood sugar levels by getting sugar into the body's cells to be used for energy or stored in the liver for later. There are two main types of diabetes. Type 1 is usually diagnosed in childhood, though it can sometimes develop in adults, and is caused by an unwanted autoimmune response which attacks and kills off the cells which produce insulin. Our risk of developing type 1 diabetes is affected by our family history and genetics and, to some extent, body weight.[7] Type 2 diabetes is more common, and most clearly associated with body weight, especially weight gained in later life. There is no abrupt threshold: the risk is cumulative – the greater the increase in fat, the greater the risk of type 2 diabetes developing.[8]

The onset of type 2 diabetes is down to either the pancreas ceasing to produce enough insulin, or the body's cells and tissues becoming less responsive to it. Being overweight or obese has the greatest impact on the latter, sometimes called insulin resistance. How and why aren't fully understood, but several things probably happen in parallel. The body's response to insulin may be hampered directly by the molecules produced by fat cells. At the same time, the inflammation caused by excess fat can stop insulin being as effective as it should be. A negative feedback loop can build up: the body's cells become resistant to insulin and absorb less sugar, which leads to surplus sugar staying in circulation. This gets converted to more fat, which in turn triggers more inflammation, so the body's cells become yet more resistant to insulin, and so on.

The symptoms of type 2 diabetes can sometimes be reversed by losing weight.[9] The disease is not cured – the underlying problem remains and can flare up again – but blood sugar levels can return to a healthy range, and symptoms go into remission. We don't really understand exactly what happens when weight loss reverses symptoms, but there is evidence that losing excess fat in the liver and pancreas in particular allows our control of sugar levels to normalise, as long as the disease has not persisted for too long.[10]

One clinical trial gave people with type 2 diabetes a low-calorie diet for twelve weeks, followed by a weight-loss diet plan. Two years later and about a third of them were in remission.[11] Three years further on, and about a quarter of those initially in remission had remained so. These few who stayed free of diabetes maintained an average weight loss of around 9 kg since the start, and no longer needed any diabetes medication.[12] But most slipped back . . . So in practical terms, reversing diabetes through weight loss is hard: it doesn't always work, and when it does is difficult to maintain.

In June 2023, a woman in China was injected with a brand-new therapy for diabetes.[13] Stem cells had been isolated from her body (from fat tissue specifically), and treated in a lab dish to become insulin-producing cells. About 1.5 million were then injected back, and two and a half months later she was producing enough insulin to stop using injections. A year later, she was still producing her own insulin. This type of therapy is not widely available yet, but many researchers and drug companies are on the case.[14] Still, while it halts the symptoms of diabetes by fixing the problem of insulin production, it won't tackle the underlying issue of inflammation in obesity-related diabetes. A variant of this therapy could deal with obesity itself by injecting cells into fat which have been modified to use lots of energy, effectively mopping up any excess sugar,[15] but for now this too is not ready to be deployed. The overall message remains simple: being overweight increases our risk of type 2 diabetes. Fact.

What about cancer?

Being overweight or obese also increases the risk of some types of cancer, including breast, bowel, pancreatic, kidney and liver cancer. Surprisingly, though, there's one example where the opposite is true: being overweight correlates with a *reduced* likelihood of developing lung cancer. This seems to imply that being overweight protects against lung cancer but, as we've seen elsewhere, correlations are not definitive proof of a causal link. In this case, what's really going on is that non-smokers tend to be heavier than smokers, and non-smokers have less risk of lung cancer.[16]

There could be several reasons for obesity more generally causing a higher risk of cancer, including hormone changes, secretions from fat cells, disruption to the gut microbiome and inflammation. For some types of cancer, like cancer in the inner lining of the womb or endometrium, specific processes are involved. Fat cells produce the hormone oestrogen, and so a high level of body fat tends to increase a woman's oestrogen level, which makes the lining of the womb grow and thicken, as happens in preparation for pregnancy. But higher than normal levels of oestrogen can also contribute to excessive growth and increased risk of endometrial cancer. To be clear, not all cases of endometrial cancer involve oestrogen, but about 85 per cent do. And around 50 per cent have been attributed to obesity.[17] In the US, in general, deaths caused by cancer are declining because of behavioural changes, screenings, diagnostics and treatments, but endometrial cancer is one of the few types increasing in prevalence and causing more deaths there every year, almost certainly because of its relationship to obesity.[18]

THE EVIDENCE OF WEIGHT

So why is our weight linked to our risk of developing other types of cancer? Recently, an international team led by Princeton University Professor Lydia Lynch set out to answer this question, and her results deserve our attention if we really want to understand the relationship between our body weight and our immune health, especially when it comes to cancer.

Lydia Lynch's team used mice, as their diet could be varied without their medical history or upbringing being complicating factors. The mice were split into three groups: one was fed normal food, another a high-fat diet for a week, and a third a high-fat diet for eight weeks.[19] After just one week, it was evident that some immune cells, called natural killer or NK cells, were being affected, in that some gene activity was modified in them. As their name suggests, natural killer cells are very good at killing diseased cells, such as cancer cells or cells infected with a virus. By eight weeks, the effect on these immune cells was seismic. A high-fat diet had dramatically reprogrammed the animal's NK cells, changing the activity of thousands of their genes. Especially affected were the genes that enable NK cells to kill diseased cells.

To test the significance of this, mice were deliberately given tumours by an injection of cancer cells, and then an injection of NK cells near their tumours. Some mice were given normal NK cells, others cells which had previously been bathed in fat molecules.[20] In those given normal NK cells the size of the tumours significantly reduced, but in mice given an injection of fat-exposed NK cells the tumours were unaffected. On top of this, NK cells normally summon other immune cells to help attack tumours, but this process was also hampered by fat. All this fits the theory that an excessive level of fat renders NK cells worse at killing cancer cells. Killer T cells – other immune cells able to kill cancer cells – also find it harder to access and kill tumours in mice fed a high-fat diet.[21]

Obesity in humans has also been shown to affect immune cells: the number of NK cells in the blood tends to be reduced,[22] and those that are present contain less of the toxic protein molecules these immune cells use to kill cancer cells.[23] In a lab dish, NK cells isolated from blood samples from people with obesity kill fewer cancer cells compared to cells from people who are lean. What's more, as with the experiments on mice, if we add fat molecules to NK cells isolated from people who are lean, we see that these cells' ability to kill is weakened.

What exactly happens to an NK cell bathed in fat molecules? Normally, there are several stages in how an NK cell kills a cancer cell.[24] First, the NK cell has to identify another cell as cancerous, which it does, as it brushes past, by the receptor protein molecules protruding from it locking on, triggering the NK to stop in its tracks, flatten up and stick tightly to the cancer cell. Then, over the next few minutes or so, small packets inside the NK cell that contain toxic protein molecules move towards the cancer cell, open up and release their deadly contents to kill it.

If the NK cells are bathed in excess fat, however, at the point when they should move the packets of toxic proteins towards the cancer cell, it's as though their innards are paralysed and they can't shoot to kill, making them less likely to deliver the lethal hit.

We shouldn't assume that immune cells in people who are lean can fight off a developing tumour easily. Irrespective of a person's weight, there are many reasons why cancer is a disease hard for the immune system to fight.[25] In some patients, unfortunately, cancer is almost, if not entirely, invisible to their immune system. These so-called 'cold' tumours (a tumour more open to attack by the immune system is often called a 'hot' tumour) are difficult for immune cells to penetrate, added to which the cancer cells release protein molecules which switch off an immune response. In fact, cancer cells release all sorts of chemicals, enzymes, genetic material and complex packets of fat, protein and genetic material we don't yet fully understand – at least some of which is certain to affect the immune system. There's a lot going on when the human body fights cancer; obesity is just one factor.

There's also an unexpected plot twist, to appreciate which we need to look further into a type of cancer treatment we came across earlier called checkpoint inhibitors (or immune checkpoint blockade), which rely on a fundamental aspect of the immune system: it's brakes.

When the body is threatened by a viral infection or other type of germ, the specific immune cells able to fight it multiply in number. A few hundred T cells can quickly become many millions, expanding the minority with the ability to see exactly which type of germ is causing the problem. A fever may be triggered, slowing down a virus and making it harder to replicate, while also helping the immune system get to work. But once the germ has been eradicated, the body shouldn't remain in this heightened state. The immune system needs to power down, the body return to its normal resting state – no more of those particular T cells are needed now – and any fever quenched. For this, the immune system has brakes.[26]

But should immune system's brakes come on during a long battle with cancer, this can work against us, slowing down the desired immune attack and allowing the cancer to prosper. In fact, cancer often *encourages* the immune system to put the brakes on. In general, the immune system's brakes are turned on when certain receptor protein molecules on immune cells bind to partner proteins on the surface of other cells, sending a switch-off message to the immune cell. Cancer cells often deliberately display proteins on the surface which can bind to brake receptor proteins on immune cells to make the immune system wind down its attack.

The receptor protein molecules that send a switch-off signal to immune cells are sometimes called checkpoints, because they also serve a subtly different purpose. Besides getting the immune system to wind down after a while, they work more generally to stop immune cells attacking the body's own healthy cells. For example, cells can be stressed or damaged if they've been exposed to a high level of sunlight, but this doesn't necessarily warrant a full-scale immune response. In this case, although there are protein molecules on the surface of cells signifying that they are damaged, as long as there are no other indications of a problem, the brake or checkpoint receptor proteins prevent an immune response. Otherwise the immune system would over-react to all sorts of things, leading to autoimmune disease.[27] (Indeed, mice that have been genetically altered to lack just one of the brakes or checkpoint receptor proteins end up dying at a young age because an enormous expansion of immune cells leads to toxic levels of inflammation.[28]) Switching off an immune response is therefore just as vital as switching it on.[29]

As a cancer treatment, checkpoint inhibitors block the negative-signalling receptor proteins from seeing their partner proteins on other cells.[30] This takes the brakes off and unleashes an immune response. Predictably, there can be side-effects from taking the brakes off, including unwanted inflammation of the skin, colon, liver or other organs. If this arises, a patient might be given counter-balancing drugs to dampen immune responses. (Occasionally, however, the side-effects of checkpoint inhibitors can be life threatening.) Such treatment certainly doesn't benefit everyone, but some people told they have a short time left have been lucky enough to survive for many years because of checkpoint inhibitors.

The unexpected twist is that this type of therapy might actually work *better* for people with obesity. A report published in 2018 documented 2,000 melanoma patients and found that the chance of survival following treatment with checkpoint inhibitors was better for people with obesity.[31] A later analysis of patients with non-small cell lung cancer also found a high body-mass index associated with checkpoint inhibitors working well.[32] A 2020 report examined over 5,000 patients with different types of cancer, and once again a high body-mass index increased the likelihood of checkpoint inhibitors working well.[33] This has come to be called the 'obesity paradox': while a high body weight increases the risk of cancer, it can also improve the likelihood of that cancer being successfully treated with checkpoint inhibitor therapy. The practical outcome of this is that when doctors have to choose which course of cancer treatment, a patient's weight could be something to take into account.[34]

It's worth noting that this isn't the case for all cancer types. One study of patients with renal (kidney) cancer, for example, found that a high body-mass index was detrimental to checkpoint inhibitors working.[35] Another, controversial factor is that the first, 2018, study also found that obesity helped men do well with checkpoint inhibitors, but not women. This variation isn't easily explained, and is yet to be borne out in subsequent studies.

The biggest question, however, is *how* obesity could improve the odds of checkpoint inhibitors working. It could be that obesity causes a background inflammation which makes some immune cells especially potent, but these immune cells are being held back by their brakes being on. Lifting the restriction, therefore, would unleash an especially powerful anti-cancer attack. Alternatively, the type of cancer that develops in people with obesity may be more likely to be 'hot' and open to attack. In other words, checkpoint inhibitors may unleash a similar immune response in people whatever their body-mass index, but cancer might be more susceptible to the boosted immune attack in people with obesity. Another possibility is that something in the microbiome of people with obesity may favour checkpoint inhibitor therapy. Essentially, anything which correlates with obesity could be involved, from differences in sleep, diet or well-being to levels of exercise.

Whatever their body-mass index, many cancer patients don't respond to checkpoint inhibitors at all (neither does obesity benefit patients receiving other kinds of treatment such as chemotherapy). But unravelling when and why obesity helps, could yield important clues to why this therapy works well, and in turn, this could help more people benefit from checkpoint inhibitors to survive cancer. Obesity is part of the disease, but might also be part of the cure.

Does being overweight make it harder to fight off infections?

During the Covid-19 pandemic, obesity was quickly identified as a risk factor for severe symptoms. This was robustly confirmed in 2022 with an analysis of over 3.5 million people across thirty-two countries.[36] Being overweight increased the risk of hospitalisation but not death, while obesity increased the risk of hospitalisation *and* death. Severe obesity also correlated with a sevenfold increase in the likelihood of invasive mechanical ventilation being needed after admission to an intensive treatment unit.[37] There are many possible explanations. One is that obesity correlates with other problems such as type 2 diabetes or hypertension, although analysis showed weight to be an aggravating factor independent of this. Age also plays a role, as young people are less likely to suffer bad symptoms with Covid-19, but obesity increases the risk in young people too. Another possibility is that body fat is a reservoir for the virus itself, but there is little evidence for this so far. What seems most likely is that obesity influences how we fare with Covid-19 because obesity itself directly affects immune health.

There are two main ways in which being overweight is thought to affect our defence against infections. First, fat cells produce molecules that trigger a background inflammation in the body, which in turn reduces the sensitivity of the system to a real threat like an actual viral infection. Secondly, in people who are obese or overweight, negative-signalling molecules are also produced by the body to counter-act the unwanted background inflammation caused by excess fat, by dialling down the immune system. But their presence then works against us when an immune response is really needed against an actual infection. The compound effect is that an immune response to some types of infection is weakened.[38]

In the spring of 2009, a new version of flu virus was detected around the world. Once again obesity increased the likelihood of hospitalisation, intensive care unit admission and death.[39] Both Covid-19 and flu affect the lungs, and part of the problem is down to the fact that obesity affects lung mechanics: excessive fat can physically limit lung expansion or narrow air passages.[40] So people with obesity generally have worse lung health, which doesn't help when an infection takes hold. This is not irreversible: a twenty-year-long study which followed over 3,000 people found that those who lost weight, improved their lung health.[41]

Tuberculosis, or TB, is another infection which affects our lungs. Although caused by bacteria rather than a virus, it also spreads by tiny droplets in the air from an infected person's coughs or sneezes. Most people don't feel any differently with a TB infection because the immune system is able to control it effectively, but even without any sign of illness, TB-causing bacteria can stay in the body for a long time – a so-called latent infection – and problems may develop later, especially if a person's immune system weakens. For example, there's an increased risk of TB reactivating if someone takes anti-inflammatory medicines or is infected with HIV.[42] Surprisingly, however, many studies have shown that being overweight or obese seems to correlate with *less* likelihood of latent TB becoming active.[43] We don't really understand why. It's particularly paradoxical because obesity is a risk factor for type 2 diabetes, and type 2 diabetes is a risk factor for active TB, from which it should follow that obesity also increases the risk of active TB. But the opposite is true.[44] What it means is that for some reason obesity helps with TB more than diabetes causes problems with it. Lots of things happening in parallel; nothing is simple.

We know that different kinds of immune response are needed against different kinds of germs, so perhaps it shouldn't be too surprising that obesity can affect the immune system in a way that can be bad for many kinds of infection but not all. What hampers an immune response against flu might have a positive effect during a TB infection.[45]

What could be a spur for action is that people with excess fat tend to respond less well to vaccines. Clinical trials have established that the levels of protective antibody produced by Covid-19 vaccines are lower, on average, in individuals with obesity.[46] This isn't something specific to these new vaccines, as responses to vaccines for flu, hepatitis and rabies are also poorer.[47] For Covid-19 vaccines, there's some evidence that protection may decline faster in people with obesity.[48] So, it's possible, but not proven, that people with obesity might benefit from more frequent booster doses.

Does being underweight also impact our immune health?

The graph comparing our risk of infections with body weight is U-shaped. The bottoming out of the U shows that a so-called 'normal' weight is associated with the lowest infection risk.[49] Being *either* underweight or overweight increases it. One reason that being underweight affects us too is simply that not getting enough food makes us deficient in all sorts of important nutrients such as minerals, vitamins, amino acids, cholesterol and fatty acids. This is why some scientists caution against only consuming diet or fat-free foods and drinks, because that could also limit the fat-soluble vitamins and nutrients we get.[50]

During a period of fasting, we also know that the human body switches into a state of limiting its energy use, which has consequences for immune health. In both mice and people, fasting for a day causes a type of immune cell called a monocyte to reduce in number in the blood.[51] In mice, these cells have been shown to move into bone marrow, where they effectively hibernate to conserve energy.[52] Eating again sees these immune cells mobilise back into the blood. In short, things definitely happen to the immune system during fasting and refeeding, but we don't yet know what effect this has on immune health or the symptoms of immune-related diseases.[53]

Intriguingly, there's a higher risk of eating disorders, including anorexia and bulimia, for children and adolescents with autoimmune diseases.[54] It is possible something in how the immune system regulates our sense of hunger and appetite could be connected to some cases of eating problems. Again, this is not well enough understood yet to enable medical intervention.

While I was writing this book, my father died. In his final months he lost a lot of weight. How ironic. Unintentional weight loss late in life is something we must certainly guard against.

So how should we think about weight?

Some scientific knowledge, like that of black holes or underwater life forms, is awe-inspiring because it shows the universe to be so much grander than we could ever imagine. The science of body weight also reveals nature to be fantastical and complex, but this is a frontier of science like no other in the extent to which it challenges us personally. The preoccupation with body weight – psychologically, physically, practically – has become almost constant in how we see ourselves and others. And yet scientifically there's so much we don't understand – about obesity, and indeed malnutrition – especially when it comes to immune health. In the meantime, in experiment after experiment we're digging away, one molecule at a time, and all manner of new human biology is being revealed.

What's the endgame? Already, a drug developed initially as a treatment for type 2 diabetes can be given as weekly injections to suppress appetite.[55] The scientific name of this drug is semaglutide, with Ozempic and Wegovy its brand names. It works by mimicking a hormone called glucagon-like peptide-1 or GLP-1 to make us feel full. This doesn't help everyone, but can lead to sizeable weight loss; in a trial of just over 1,600 people with obesity or overweight, those taking weekly injections lost an average of nearly 15 per cent of their weight within sixty-eight weeks, compared to the placebo-control group losing just over 2 per cent.[56] There are problems with this medication, though – not least that this kind of medicine can become a life-long commitment, because stopping leads to considerable weight gain. There can be side effects too, such as nausea, diarrhoea and vomiting. There are supply problems too: people with type 2 diabetes are already finding it hard to obtain the medicine – for its originally intended use – because of the demand for weight loss. Still, it is surely only a matter of time before new drugs will be developed that can be taken orally, are easier to produce and work even better. We don't yet know the extent to which such drugs would fix the many different health problems associated with obesity. But there is mounting evidence that GLP-1 drugs could benefit all sorts of conditions from heart disease to Parkinson's.

There are other issues, such as who will pay for medicine like this, given the vast number of people who could benefit, and especially if this needs to be given forever? In the UK, this is often decided by comparing the cost with how much money can be saved by preventing other problems. But that's not all. For example, the widespread availability of drugs for weight loss might stop us tackling the root causes of global obesity. And at what age should we begin to medicate against overweight or obesity – middle-age, young adults or children? This is not a hypothetical question. In 2022, a small trial showed that injections of a GLP-1 mimic can help teenagers lose weight.[57]

We know that keeping at a healthy weight is important for many reasons, including immune health, which controls our susceptibility to infections, cancer and other illnesses. This is because many immune cells reside within body fat itself, and because lots of molecules released by fat cells directly affect the immune system. But though levels of fat in the body clearly impact our immune health, science can't yet be clear as to what we should do about it. There's no 'one-size-fits-all' approach to maintaining a healthy weight. Each of us must decide for ourselves our attitude to weight, psychologically, physically and practically.

4

The Yin and Yang of Exercise – or, how much exercise is best?

Many years ago, my wife and I walked into a gym to ask about joining. We were directed to talk to someone wearing a business suit, not a tracksuit. It turned out it would be a lot more expensive than we had imagined, so we said we'd think about it. As the salesperson opened the glass door for us to leave, he nudged me gently to look at all the cars in the car park. It must have been obvious I didn't understand what he was getting at. 'You'll want to be a member of this place,' he said, 'because we have people who drive these kinds of cars.'

Of course, there's nothing wrong with joining a gym to socialise, go on dates or seek new business contacts. Also, health is not merely the absence of an illness, but a state of physical and mental wellbeing for which social interactions are important, and in that sense a gym may indeed benefit our health beyond providing equipment for exercise. But fitness clubs will try almost anything to get us signed up. It works in their favour that there's a superiority associated with fitness and slenderness, disconnected from what it takes to be healthy. Looks are judged. Fashion plays a role. Social media can be mean. All over the world, an enormous amount of money is spent on exercise and sport, from shoes to treadmills. Once again, we are bombarded with messages – adverts, social media posts, magazine articles and podcasts – that leave us perpetually feeling guilty, this time, about not exercising enough. It's important to step back from this, to take stock of what we really know. Scientific scrutiny – measurements of immune cell numbers and activity before, during and after exercise, and population-level studies – helps us think about the appropriate levels of exercise for our own lives.

It's been clear for decades that physical activity reduces the likelihood of all sorts of medical problems, including cancer, cardiovascular disease, type 2 diabetes, chronic inflammatory disorders and stroke. One landmark study published in 1986 followed nearly 17,000 Harvard alumni for between twelve and sixteen years. Those who played sport, walked regularly or undertook some other exercise were less likely to die of *any* cause.[1]

More recent research has discovered that a physically active lifestyle correlates with a reduced risk of developing breast, colon, endometrial, bladder, stomach and kidney cancer by 10 to 20 per cent.[2] There's also evidence that people who regularly exercise have a better chance of survival after being diagnosed, at least for breast, colon or prostate cancer. Irrefutably, exercise is important. This much is drummed into us from school onwards. But recent advances in science can now shed light on further things: how much exercise we should do for immune health, whether you can do too much, and what exactly happens to the immune system when we exercise. First, though, a couple of myths to bust.

How much exercise should I do for immune health?

One common belief is that we need to exercise more these days because we've become couch potatoes. For most of human history, goes the argument, we've been hunter-gatherers rather than binge-watchers of TV shows, and the human body has adapted accordingly – across thousands of years of evolution – to work well with high levels of physical activity. This profound insight into the human body has one big problem: we don't know if it's true.

Do we really sit around resting, watching or contemplating more than our ancestors? Surely people have always relaxed, had time for chatting, gossiping and hanging out? According to fossils, Palaeolithic populations commonly knelt or squatted, presumably to relax.[3] But it's true they didn't have chairs. Chairs with backs and armrests date to around 5,000 years ago (and were used as a status symbol by the ancient Egyptians). Nowadays sitting is sometimes described as 'the new smoking'.

There is a correlation between the length of time a person sits down each day and an increased risk of many medical problems, including heart disease, type 2 diabetes and cancer. Though such health risks shouldn't be discounted, this doesn't prove one thing causes the other. If sitting was the problem, then sitting at work versus sitting at home should have the same negative consequence for health – and yet there is evidence that sitting at home is worse for our health.[4] Perhaps something else is at play, then, like socioeconomic factors.[5] A study following over 80,000 people for nearly seven years showed that people using a standing desk did not have less risk of developing cardiovascular disease (in this study, immune health was not looked at).[6] So it seems unlikely that there's some ancient exercise regime the human body has adapted to which we must hark back to, or that, because of how we used to live, modern levels of sitting *must be* causing problems with immune health in the general population.

While sitting versus standing is still a matter of debate,

exercise certainly matters. To stay healthy, the UK Chief Medical Officer recommends that adults take at least 150 minutes of moderate activity each week, like brisk walking, dancing or hiking, or 75 minutes of a more vigorous activity: running, swimming, riding a bike fast. Other countries have similar if not identical guidelines, though advice varies by age in some places. In the US, for example, young people are recommended to take 60 minutes or more of moderate-to-vigorous activity, and older adults are advised to include balance exercises like walking backwards, standing on one leg or t'ai chi.[7] Such guidelines are not wrong, but can be misleading, especially in relation to immune health.

A target is useful because it is motivating, but in truth there is no threshold level of exercise that's especially beneficial.[8] It is a continuum. So, guidelines could be taken to imply that doing less than the recommended level of exercise is not good enough, which is not true. For most people it would generally be more accurate to say that there's a benefit to becoming more physically active. (The exception, for extremely challenging exercises, we'll come to later.) And the greatest boost in health – not only immune health but also protection from all sorts of diseases including heart problems, cancer, type 2 diabetes and stroke – comes to anyone starting routine physical activity from previously doing little or none. For most of us, any level of exercise is good for you, and any increase in the amount you do, is also good.

What happens to our immune system when we exercise?

In 1902, it was noted that runners in the Boston Marathon showed an increase in the number of immune cells circulating in their blood, and we've been studying this ever since.[9] We now know that almost immediately after even a moderate bout of aerobic exercise, immune cells are mobilised to move out from bone marrow to circulate in the blood.[10] We also know that increased blood pressure, and just the actual movement of the body itself, boosts the number of immune cells flowing in blood by detaching them from the inner margins of blood vessels.[11] Potentially, exercise could alter a person's gut microbiome, which may also affect the immune system (as we've already seen in Chapter 2).[12]

One type of immune cell increased in the blood by exercise are neutrophils. These are normally the most abundant immune cell anyway in the blood, but exercise swells their numbers. One of their jobs in the body is to gather at a cut or wound within minutes to engulf opportunistic bacteria and destroy them. They can also create web-like structures made from strands of DNA and proteins to capture, kill or neutralise bacteria, fungi, viruses and parasites.[13] More of these cells flowing in your blood sounds a good thing – protecting you better against any opportunist germs, for example. But some studies have suggested that during intensive exercise, neutrophils can release their web-like structures even in the absence of any germs. It's not clear why this happens, but it has been speculated that in athletes this could contribute to muscle or tissue damage.[14]

The activity of muscles is also directly connected to the immune system, because of all the different hormone-like molecules muscles produce. As well as enabling movement, muscle can also act a little like a gland, sending out molecules that affect all sorts of other tissues and organs throughout the body. Astonishingly, around 600 different types of protein molecules can be released by muscle when it contracts.[15]

Collectively they are called myokines, and they are sensed by receptor proteins present on all sorts of other cells in the body, including those in the liver, pancreas, bones and brain. One myokine, called irisin, crosses the blood–brain barrier and directly acts on brain cells. Although little is clearly established about what this means for health, it has been suggested that irisin from muscle can act as an anti-depressant, and even protect against Alzheimer's disease.[16] This is a significant frontier: there's much to be discovered about myokines linking exercise and mental health. It's possible that one day myokines, or mimics of myokines, or the receptor protein molecules on the brain cells which myokines act on, may become the basis of new treatments for mental health conditions.

Myokines also act on the immune system. One, called IL-6, is increased in the blood by up to 100 times whenever we exercise, declining around an hour after we stop. IL-6 is very important in immune responses, involved in all sorts of things including antibody production.[17] It would be reasonable to conclude that this indicates an important link between exercise and immune health, were it not that immune cells also produce IL-6 during an infection. So although this myokine has long been considered a potent activator of immune responses, this might only be in the context of an infection and when it's released by immune cells. The specific effect of IL-6 coming from muscle is not entirely clear[18] – in fact, there is some evidence that, in the absence of an infection, IL-6 from muscle acts to quieten down the immune system, which might be detrimental if a person is then exposed to an infection, but beneficial if a person suffers from an unwanted inflammation such as rheumatoid arthritis.

This sort of thing comes up a lot in the science of exercise and the immune system. There's good evidence for a specific change in the immune system – a change in blood neutrophil numbers or the amount of IL-6 produced by muscle – but less for what its exact health impact is. So what does exercise really do for immune health in terms of preventing disease? Let's start with cancer.

Can exercise help fight or even prevent cancer?

The first suggestion that muscular activity might protect against cancer came from the observation that farmers who were physically active in their seventies and eighties tended to have a lower risk of developing cancer.[19] We now know to be careful of correlations like this, but nevertheless we do know that exercise has the immediate effect of increasing the blood numbers of natural killer (NK) cells, for example – the immune cells able to directly kill cancer cells as well as some types of virus-infected cell. NK cells detect protein molecules on the surface of cancer cells that mark them out as cancerous because they are seldom found on healthy cells. If this happens, an NK cell will latch onto a cancer cell, flatten up against it, kill it, detach from the debris and then move on to attack another. And there are at least two ways in which exercise helps in this. First, with exercise, individual NK cells become more efficient at this process (precisely why we don't know). Secondly, a sub-type of NK cell particularly good at killing cancer cells is especially elevated in blood after exercise. Both effects increase the chance of a kill when the NK cell meets a cancer cell.[20]

All this helps our understanding of how exercise *could* reduce the risk of some types of cancer,[21] but it's hard to test this in humans, partly because cancer is complex and everyone's situation is so individual. So scientists have turned to mice, with whom the timing and duration of exercise, and their exposure to cancer, can all be controlled. What's more, the basic biology of mice and humans are, perhaps surprisingly, remarkably similar, which means they too have NK cells that are very good at killing cancer cells.

One hugely important and influential experiment went as follows. Mice were split into two groups: some caged with a running wheel, others not.[22] Mice are running enthusiasts, and those with a wheel voluntarily ran about 4–6 km a day. Both sets of mice – running and non-running – were then injected with cancer cells and monitored for the development of tumours. The number of tumours can be taken as indication to how well the mouse immune system could fight off a developing cancer. Mice given a running wheel *after* being injected with cancer cells fared the same as if they had no exercise. But mice running for four weeks *before* being exposed to cancer cells benefitted to a shocking extent: they developed less than half the number of tumours as non-running mice.

In another version of this experiment, instead of mice being injected with cancer cells, they were injected with a chemical that caused tumours to develop. MRI scans over the following eleven months showed that 75 per cent of non-running mice developed tumours in their liver, compared with only 31 per cent of running mice. Another dramatic consequence of running.

In this case, giving mice a wheel *after* they were injected also worked to limit tumours developing. One explanation for this difference from the last experiment, where only running beforehand helped, is that the chemical causes tumours

THE YIN AND YANG OF EXERCISE

to develop over a much longer period – many months, in fact – while an injection of cancer cells is an almost instant onslaught. So this result could be explained by exercise taking some time to benefit the immune system: too slow to control an injection of cancer cells, but quick enough to help stop cancer when it's developing more slowly.

To try and understand how exactly running was helping, the team looked at what was happening inside the tumours. Inside tumours that did develop in running mice, there was an infiltration of NK cells, as though these immune cells were at least trying to keep the cancer under control. By contrast, in non-running mice NK cells were seldom if ever found in tumours, which fits with the notion that an increased number of NK cells, and an improvement in their ability to work, is pivotal to how running, and exercise more generally, could help fight cancer. To test this specifically, mice with cancer were treated with a drug that kills off their NK cells. Lo and behold, in mice lacking NK cells the benefit of running was lost.

More evidence for the importance of NK cells comes from experiments blocking the action of adrenaline. This can be done by giving mice a drug called propranolol (used to treat heart problems and help with anxiety). Amazingly, blocking the actions of adrenaline in running mice stopped NK cells reaching tumours. What's more, this was found to work the other way around too, by giving adrenaline to mice that weren't otherwise getting a hit from running. If non-running mice were injected with a low dose of adrenaline every day for one to two weeks, their NK cells were more frequently found in tumours. So a low dose of adrenaline, like that induced by exercise, helps these cancer-killing immune cells get to where they're needed.

We know that NK cells are good at killing cancer cells, so this all fits with how exercise can help fight cancer. But there's something else going on. Research my own lab team played a role in, led by Santiago Zelenay at Cancer Research UK's Manchester Institute, has examined the sequence of events when immune cells attack a tumour.[23] It's very important that NK cells arrive at the tumour early in its development. But as well as killing cancer cells, NK cells produce protein molecules which act as a bat-signal or call to arms to summon other types of immune cell.[24] In fact, the ability of NK cells to announce the presence of cancer to other immune cells may be their most important contribution to how our body can sometimes fight cancer.[25] A signal from NK cells calls in a cavalry of all kinds of immune cells into a tumour, including killer T cells.

Again, we can use mice to see if and how exercise could affect killer T cells.[26] In this experiment, mice were split into two groups, one given a spinning wheel to run on, others not. The non-running mice were also given a spinning wheel, but one that didn't turn. After fourteen days, both sets of mice were given an injection of breast cancer cells. Running mice developed fewer tumours and lived longer, although most died in the end. But crucially, it was removing killer T cells that wiped out the positive effect of exercise.[27] More dramatically, isolating killer T cells from running mice and injecting them into less active mice helped protect these mice from the development of tumours.[28]

So in some experiments NK cells are vital, in others killer T cells; it probably depends on the type of cancer. And exercise itself has multiple ways of affecting the immune system. We've seen that adrenaline is important, because blocking it reduced the number of NK cells entering tumours. But experiments where killer T cells were more important highlighted a role for something else: a small chemical called lactate. Lactate is produced as by-product from muscle cells breaking down carbohydrate for energy. During exercise, lactate increases in the blood and, at least in mice, boosts the ability of killer T cells to attack cancer. Daily infusions of lactate increased the number of killer T cells which infiltrated tumours to improve an animal's chance of survival.

Increased numbers of immune cells in the blood and inside tumours would seem to be a good thing. But it's important to be aware of the good and bad things that can happen when immune cells infiltrate a tumour. Sometimes immune cells get inside it and destroy it – the outcome we want. But sometimes an immune response can have the opposite effect and actually help the tumour prosper. As far as this is understood one line of thinking runs as follows.

We know that as well as defending our body against germs, immune cells can be involved in wound healing. At a cut or site of infection where the tissue has been damaged, immune cells produce growth factors which encourage local cells to multiply and repair the damage. So in certain situations, it is possible that tumours like to engage with immune cells because they can trigger immune cells to secrete growth factors which benefit the growth of the tumour.

Thankfully, exercise seems to promote the right kind of immune response. In mice given pancreatic cancer, exercise increased cancer-attacking immune cells within tumours, and lowered the abundance of other types of immune cell the tumour could benefit from.[29] A small trial involving people with pancreatic cancer had patients exercise for two hours a week (one hour of aerobic activity and one hour of strength training) before surgery was performed to remove their tumours.[30] The tumours were later sliced up and studied under a microscope. Those who exercised had significantly more killer T cells present in their tumours, and their killer T cells contained higher levels of the toxic protein molecules they use to kill cancer cells. Exercise, then, doesn't only increase immune cell numbers in a tumour but also seems to get the right type of cells there. All in all it seems that exercise can, in myriad ways, help our immune system fight cancer.

Does exercise help us fight infections?

As we know all too well by now, a respiratory virus like that which causes Covid-19 moves between us in droplets and is spread through sneezing, coughing and talking. Germs can also spread by lingering on surfaces: the flu virus can stay on a hard surface for twenty-four to forty-eight hours; some bacteria like *E. coli* probably survive a few hours; others, like *C. diff*, can survive for a few months outside the body.[31] Fungal spores spread by person-to-person contact or in the air. Exercise can't affect any of these processes, to stop an infectious agent entering the body. So for infectious diseases, the question is this: does exercise help us fight germs once they've entered the body? There are two effects to consider.

First, exercise or anything else might help our immune system deal with a germ very quickly after it enters the body, in which case our immune system would eradicate the germ before we even are aware of it. This is often taken as a reduction in the risk of infection but, to be clear, a germ still gets into the body; it's just that it's fought off so quickly it's as if nothing has happened. Secondly, exercise or anything else might help the immune system fight off germs that weren't killed off quickly and have multiplied and spread around the body. This might involve a fever and us feeling ill for a while. In this case, if exercise helped, it would reduce the severity of symptoms, or the time we are ill for.

Evidence that exercise helps reduce the risk of an infection comes from a year-long study of almost 19,000 people living in Denmark. Across this huge group, it was established that any level of physical activity correlated with a person being less likely to need a prescription for antibiotics.[32] This overall trend was heavily influenced by women who exercised being much less likely to report a suspected urinary tract infection. Other studies have found that people who exercise tend to experience lower rates of upper respiratory tract infections including a cold.[33] One large study in China, for instance, found that exercising moderately for three or more days per week was associated with 26 per cent less risk of having at least one cold a year.[34] A UK study of nearly 100,000 people correlated exercise with halving the risk of dying from an infection over a nine-year period.[35]

It's important to say that these correlative-based findings are not definitive – there could be any number of things that vary between people who exercise and people who don't – but they fit with the notion of exercise helping the immune system kill off germs quickly, either to the extent that it doesn't become a fully-fledged infection, or severe symptoms are less likely.

To test this directly, small, randomised, controlled trials have been carried out in which people are split into two groups, one with an exercise regime to follow, the other not. To give an example, one study followed 149 people of average age sixty. An exercise plan consisting of a weekly

2.5 hour group session plus forty-five minutes' daily home workout for eight weeks led to people having 43 per cent fewer days off sick during one cold and flu season.[36] The research team also undertook a slightly bigger study, with 390 people involved, and found a 21 per cent decrease in the number of days off work for those exercising.[37] The second trial involved younger people, average age fifty, so it's speculated that the smaller effect was down to the importance age plays in immune health (as we'll see later). Most importantly of all, however, neither of these results reached statistical significance – the number of people involved was just too small. Arguably I shouldn't even mention results which haven't been proven to a statistically significant level. On the other hand, this is what we know and don't know, unvarnished and as messy as it is. As interesting as this all sounds, larger studies are needed before this research should be taken to mean anything practical.

Again, research with mice is more easily controlled, and by being more easily controlled, statistical significance is much easier to reach with small numbers. And the results fit with the human studies in terms of exercise protecting against infections. Mice running for forty-five minutes a day at moderate speed, five days a week over fourteen weeks saw them showing less severe symptoms when infected with flu – for example, exercise reduced their infection-associated loss of appetite.[38] Perhaps more importantly, two days after being infected, the virus multiplied to a far lesser extent in the running mice's lungs.

Some illnesses caused by a virus aren't contracted by being exposed to a germ harboured by somebody else. They come from within ourselves. That's because some viruses have a way of becoming dormant, or latent, inside the human body, only to reawaken and cause problems later in life. The virus that causes chickenpox does this.[39] After a bout of illness, the virus isn't entirely eradicated, but stays in the body, quietly hidden, for many years. It may re-activate later and cause shingles. Cytomegalovirus is another virus which stays in the body: after a first infection it is there for life. It is usually harmless, but can occasionally cause serious problems in unborn babies or people with a weakened immune system. A study of over 1,000 people found that regular exercise and increased general fitness were associated with a lesser likelihood of a latent cytomegalovirus reawakening to cause a problem.[40]

Crucially, however, there's a flip side to all these positive benefits of exercise.[41] Stress hormones can be produced during exercise and (as we will see in the discussion on stress in Chapter 5) these tend to dampen immune responses. Immune cells also need a lot of energy to be able to secrete toxins to attack diseased cells, and multiply in number by dividing. During exercise, energy is used for muscle activity, and that available to immune cells is therefore limited. Which is partly why most doctors advise not to exercise heavily during a bout of flu or in the grip of fever, because the immune system needs your body's energy to fight.

Which leads us to our next question: how does intense exercise affect immune health?

Is intense exercise better or worse for immune health?

Perhaps surprisingly, exercising more strenuously isn't inevitably better for immune health. At a certain point there are detrimental effects too, at least in the short term. Various studies of athletes, including ultramarathon and marathon runners, elite swimmers, cross-country skiers, track and field athletes, have reported a greater susceptibility to immune-related problems, especially respiratory infections.[42] For example, athletes in Team Finland at the 2018 Winter Olympics tournament were more likely to catch a cold than other staff members.[43] Ultramarathon runners are also more likely to have asthma.[44] An analysis of all the data available for triathlon participants found that inflammation markers were often increased after a race or training, and concluded that to counterbalance this athletes should pay careful attention to other aspects of their life which could help with immune health, such as nutrition and sleep.[45]

Measurements of immune cell numbers in the blood following a bout of vigorous exercise suggest there's a short-term suppression of the immune system which lasts a few hours.[46] This may allow for an opportunistic infection to occur – the so-called 'open window' theory, though this remains contentious: it may be that people who have just run a marathon race, for example, are more susceptible to an infection simply because they have just been among a huge crowd of people.[47] It's also difficult to disentangle the effects of strenuous exercise itself from the stress of competition, and an altered diet or sleep patterns. So while high levels of training and competition have been linked to worse immune health and an increased risk of infections or other illnesses, the exact reasons remain unclear.

In the long term, however, a different story emerges. In general, our immune system weakens as we age. For example, 80 to 90 per cent of those who die from the flu virus are over sixty-five, in part owing to a weaker immune response against new versions of flu.[48] Older adults also respond less well to vaccination. It's not that our immune system simply stops responding, because older adults are also more likely to suffer from an autoimmune disease, which is caused by an unwanted immune response. Rather it seems that as we age our immune system goes awry. Why isn't known for sure, but one idea is that a low level of background inflammation persists in older adults, perhaps caused by an accumulation of tissue damage. This background inflammation throws the system a little out of kilter, making the immune system worse at detecting germs and slightly more prone to attacking the body's own healthy cells. One way exercising in old age may help is that myokines like IL-6, produced by muscle activity, quieten this background level of inflammation.

A history of fitness may also help. Older runners who had trained for at least seventeen years or more had better T cell function than those who hadn't run much during their life, which fits with the principle of them being better able to fight infections.[49] Another study found that 65-to-85-year-old men who maintained a moderate or high level of exercise for around twenty-five years had better antibody responses to a flu vaccine compared to age-matched men who hadn't exercised much.[50] Another analysis found that swimmers, judo and track and field athletes who had competed for over twenty years had preserved a larger pool of T cells of the specific type equipped to counter germs the body hasn't been exposed to before.[51] Needless to say, ageing of the immune system – and ageing in general – is enormously complex and not very well understood, but the evidence suggests that exercise in old age, plus exercise throughout a person's life, is good for immune health.[52]

Does the type of exercise matter?

Most of the research looking at immune health after exercise focuses on the effects of endurance training or aerobic exercise. Relatively little has been done to compare different types of exercise, such as resistance training or weightlifting.[53] Interestingly, though, one very recently discovered process by which exercise can increase the numbers of immune cells in blood comes from the physical force it puts on our bones. Cells which line the outside of blood vessels within bone marrow directly sense mechanical pressure, and this stimulates them to produce growth factors. These growth factors trigger the formation of new bone cells to strengthen bones. They also increase the production of immune cells from stem cells in the bone marrow.[54] So the raw physical force of pressure on our bones could be a factor in increasing blood immune cells, which maybe means that different types of exercise could have subtly different consequences – say, weightlifting versus swimming, which would involve greater or lesser pressure on bones.

Does exercise help with vaccines or other medicines?

Exercise can also help with immune health by working alongside medicines. Several studies, though not all, have found that something like a forty-five-minute brisk walk or workout just before having a vaccine can improve responses.[55] But there are complications: one study found that a forty-minute, moderate-intensity aerobic exercise improved the response to a flu vaccine of women aged fifty-five to seventy-five. So far so good, but consider this: the flu vaccine is designed to protect against several different strains of the virus – but the boost from exercise only happened against one strain, H_1N_1, and not another, H_3N_2.[56] There's no obvious reason why exercise would boost a person's response depending on the strain of flu but – I'm speculating here – it could depend on the strains of flu people had previously been exposed to. Another unexpected finding was that a boost from exercise was found for elderly women and not men – also not easy to explain.

The number of people tested was very small (ten to seventeen in each group), and one important factor that was not recorded was the fitness level of participants, and whether they exercised routinely. In any case, a major problem in assessing the significance of these results is how vaccine effectiveness is measured. This is almost always done by measuring the amount of flu-attacking antibody in someone's blood sample. Antibodies are certainly vital, but it's difficult to know if or to what extent a 10, 20 or 50 per cent increase in antibody levels impacts someone's ability to fight off an actual flu infection. There haven't been many studies yet testing whether other kinds of medicine might be affected by exercise, but one indicated that running boosted the effectiveness in mice of an immune therapy against breast cancer.[57]

What can we conclude?

Many lines of evidence point in one direction: *exercise helps*, especially for cancer, some types of infection and in old age. There's no clear threshold, however, of the particular level of exercise we all need to reach. In fact, the question of how much exercise doesn't have an easy or simple answer – the scale of its effect on immune health is hard to measure, especially the variation from person to person.

The level of exertion is certainly significant, but again not as straightforward as you'd imagine. There's no simple dose-dependent relationship: short- and long-term consequences can differ, even diametrically. Extreme levels of exercise are particularly complex to assess, with their effect bundled up with that of stress on the body. Overall, there is evidence that extreme exercise can lower immune health in the short term, but then be beneficial later in life.

Sometimes, I know, we just want the bottom line: a precise answer to a question like how much exercise I should do for good immune health. Likewise, we would love to know what simple thing will make us feel better quickly whenever we're feeling ill, hence the power and seduction of drinking a glass of orange juice to help with a cold, but as we've seen, this isn't scientifically proven to help. So what we need – what is truly going to empower us – is the chance to understand the science. The opportunity to carefully consider the results of research in immune health means you can then decide for yourself what to put into practice in your everyday life.

5

A Reason for Calm – or, does stress impact immune health?

Prague, 1925. Having entered medical school the previous year, Hans Selye starts to see patients for the first time. He's a diligent, loyal student – impressed by his professors' sharp eye for diagnosis – but the more hands-on experience he gains, the more he realises his professors are missing something. Most of his patients, regardless of their diagnosis, have similar generalised complaints. They lack energy, they can't sleep, they've lost their appetite.

The professors, thinks Selye, may have become specialists in the symptoms of various diseases, but there's an elephant in the room: a syndrome of *just being sick*. He floats the idea to one professor, and is told to concentrate on his exams.[1]

Selye continues with his career like any other medic, but ten years later he returns to his theory of some symptoms being common to nearly all illnesses. He's carrying out tests for hormone activity, which involves injecting rats with extracts derived from the ovaries of slaughtered cows to see how they react. A dummy injection, he notices – one without any hormone or extract – also causes symptoms. It occurs to him that the injury itself – the fundamental unpleasantness of it – could be responsible. So he investigates further. He stations rats on a cold roof knowing they prefer being warm. He sets others on a strenuous revolving treadmill. Animal cruelty aside (this is the 1930s), what he finds is that all sorts of unpleasant things – or stress, the term he would go on to use – triggered notable responses, including an enlargement of the rats' adrenal glands and ulcers in their stomach and upper intestines.[2] At the age of just twenty-nine, Selye published a landmark article in *Nature* arguing that 'a typical syndrome appears' when animals are exposed to all sorts of stressful situations.[3] Stress, he had discovered, shows up in the body.

Nowadays we all know that stress can affect us profoundly, both physically and mentally, which begs the question as to whether stress also affects our immune health. While much of what we're discussing in this book remains up for debate, there is absolutely no question that it does.

Let's start with the fundamentals: what is stress?

A single word can cover so much, and that 'stress' covers so many situations is part of its power: sitting an exam, running to catch a train, losing a loved one, playing a sport, trying to soothe a crying baby, riding a roller coaster or moving house. Yet without a fixed, narrow definition there is also a fuzziness to what we mean by stress. Is it stress we feel watching *Jurassic Park* when water ripples in a cup because a dinosaur is stomping nearby? Surely watching *Jurassic Park* is not going to cause stress to an extent that it affects our health? 'Stress is not even necessarily bad for you,' Selye wrote. 'It is also the spice of life.'[4] Short bursts of stress can be exciting, motivating and assist performance.

We have a complex love–hate relationship with stress that permeates our lives. On the one hand there is something glamorous about being busy. We all know people who thrive on telling us how much they have going on. In TV dramas like Aaron Sorkin's *The West Wing* the characters talk while walking fast to emphasise just how packed their power-wielding lives are. But how many times have you felt that life is moving just too fast, leaving you tired and wiped out? Worse still, stress can be used as an excuse to be short with someone, not read your child a story or be aggressive towards someone you love. Countless products are marketed to help us de-stress, relax or feel calm. Evidently stress is a knotty construct born out of biology and culture, not easily unpicked, but scrutinising the science helps. We have a broad understanding of the changes in the body that occur during stress; one hormone in particular, cortisol, has a central impact on the immune system. But what also emerges is that there's much more to this story than there appears to be at first.

Are some of us more prone to stress?

Everyone experiences stress at some point. It could be problems with a relationship, financial issues, the death of a loved one, trouble at work or simply world events. And even positive things in life can be stressful: having a baby, changing job or planning a party. All of our lives follow a unique path. We also know from everyday experience that people vary in how they handle stress, which could possibly be related to personality types. In the mid-1950s, two cardiologists in the US, Meyer Friedman and Ray Rosenman, coined the term 'type A personality' to describe people who are exceptionally ambitious, competitive and impatient. People with type A personalities, they went on to suggest, lead such stressful lives that they are making themselves prone to illness. Is this true?

We do know that specific traits correlate with our response to stress. One study, for example, reported that perfectionists tend to exhibit higher cortisol levels after a mock job interview.[5] But the idea of there being a handful of personality types is simplistic. Traits like perfectionism, competitiveness and impatience occur in people along a spectrum, rather than any of us falling into yes/no boxes.[5] Overall, our response to stress is individual: factors include sex, genes and mindset, but these are extremely hard to disentangle to the point of enabling us to make predictions about how people respond to different kinds of stress. There is actually not a lot of evidence that a type A personality really exists, never mind it putting someone at greater risk of stress-related illnesses or poor immune health, but the assumption has tended to persist.

Are we more stressed today than ever before?

The idea that we're more stressed now than at any time in history is prevalent these days. Technology and advances in science seem to have forced us into the fast lane, and left us worrying that modern life has become just too stressful. Then again, our ancestors had a shorter life expectancy, more poverty and no prospect of ever having a job with holiday entitlement. In 1977, an interviewer asked Hans Selye if he thought life had become too stressful. 'People often ask me that question,' he replied. 'They forget that the caveman worried about being eaten by a bear while he was asleep, or about dying of hunger, things that few people worry about much today . . . It's not that people suffer more stress today. It's just that they think they do.'[6]

Cortisol levels have been measured in hair found in archaeological sites in Peru, from bodies that are 500–1,500 years old.[7] It turns out that they were stressed too, we can only imagine what by.[8] So overall it is probably a myth that we are more stressed today than ever before – or at least, it's hard to know for sure. All we can be certain of is that the causes of stress are different now. I imagine that stress has affected people's mental and physical health throughout history.

So how does stress impact my immune health?

After Selye, an enormous, painstaking, game-changing adventure unfolded in which tens of thousands of scientists set out to unpick what happens in the body during stress. The result was our current knowledge of the sequence of hormonal changes that take place to get the body in a heightened state for the so-called 'fight-or-flight' response – the body's reaction to a harmful event, attack or threat. A similar cascade happens whether we encounter real danger or something less threatening like being stuck in a traffic jam. The hypothalamus, located directly above the brainstem, releases a hormone called corticotropin-releasing hormone, or CRH, which acts as a signal to the pituitary gland. The pituitary is a pea-sized gland just below the hypothalamus that acts as a kind of master gland, producing many hormones that enter the bloodstream to regulate bodily functions from growth, blood pressure and energy use to temperature and pain relief. In stress, the pituitary gland sends out a hormone called adrenocorticotropic hormone, or ACTH, which is sensed by two small glands sitting on top of your kidneys, the adrenal glands. The adrenal glands then release other hormones, including two famously connected with preparing the body for action: cortisol and adrenaline (also known as epinephrine).

There are nuances and feedback loops: when cortisol levels are high, this is sensed by the brain, which then shuts down further release. If the stress is only fleeting, this negative feedback brings the body back to its normal resting state. To give a specific example: people taking a tandem skydive for the first time, an experience both stressful and with an element of real risk, show a change in their white blood cell count immediately post-jump, but this returns to their normal baseline levels within an hour.[9] This is why there are no long-term consequences from the tension of watching a movie. But if stress persists, the adrenal glands may continue to elevate cortisol and other hormones. It's in such circumstances that you might get the feeling that you are constantly running for a train.

These hormones – both cortisol and adrenaline – increase your heart rate and elevate your blood pressure. They also trigger the liver to increase blood sugar levels, to boost the energy available to cells and tissues. In this way, hormones prepare the body for a change in activity, which is important for lots of situations, not just stress – for example, a rise in cortisol in the morning helps prepare the body for waking up.[10] Nevertheless, cortisol levels change much more dramatically with stress.

As part of preparing the body for action, stress hormones shut down or quieten body systems not immediately needed. Sexual feelings are inhibited – presumably a waste of energy and potentially a major distraction during an emergency. More pertinently, immune responses are quietened too. Here the effects of cortisol have been studied the most. Receptor proteins on immune cells directly sense cortisol, and this dampens the immune system's ability to respond to germs.

Dialling down immune responses saves energy and avoids inflammation in the body's tissues that might slow us down, which is fine for a brief time, but if stress persists, our immune system can stay weakened. One way of thinking about this is that the system has evolved to help us run from a predator, and doesn't necessarily work as well for modern stresses like changing jobs or going through a complicated divorce.

Over a hundred studies have reported that long-term stress can contribute to poor health in one way or another.[11] People who are stressed for prolonged periods of time suffer worse symptoms from viral infections, take longer for wounds to heal and respond less well to vaccination.[12] Stressed individuals are more prone to the reactivation of herpes, for example.[13] People stressed by caring for a spouse with dementia have a reduced response to the flu vaccine.[14] Also, the chance of men infected with HIV developing AIDS increased two to three times over a five-and-a-half-year period, if they had higher than average stress or less social support.[15]

Stress affects animals too. Obviously an alarm for jumping into action is crucial to survival for all sorts of creatures on earth, from fish to birds to mammals.[16] Here too, stress has an effect on an animal's immune health, adding weight to the general theory that stress impacts immune health deep down and fundamentally. In one experiment, rats that were stressed by smelling the odour of a predator were less able to fight off a fungal infection.[17] In another, mice were restricted in their movement, able to move back and forth but not turn around, before being given a dose of flu. Their immune response was delayed, and fewer immune cells moved to their infected lungs.[18] Crucially, if these stressed mice were given a drug that blocked the effect of cortisol, their immune system responded normally. Also, a drug interfering with the receptor proteins that sense stress hormones reversed the effects of stress on immune health in mice.[19] These experiments and others like them provide the evidence that stress affects immune health because of hormones, especially cortisol.

Mice do not naturally get Covid-19 because they have a different version to humans of the receptor protein ACE2, which the virus uses to enter cells. This is why there was something of a scientific race, early in the pandemic, to somehow make animals susceptible to infection in order to understand the disease's symptoms and test possible treatments.[20] Several labs bred genetically altered mice to give them the human version of ACE2 that made them able to contract the virus, and recently these modified mice were used to test the effects of stress on Covid-19. Two groups were injected with the virus, one free to run around in their cages as normal, the other stressed by being restrained for four hours. Over the following days this second group was restrained three more times, to mimic continual stress. One week after infection, the lungs of the stressed mice contained much higher levels of the virus. Five days later, the stressed mice were far more likely to have died.[21]

Studies of mice are important in revealing details of *how* stress affects immune health.[22] Normally, immune cells are constantly on the move, as it is vital for them to patrol the body in order to be able to get to the site of an infection, and for different types of immune cell to interact with one another. But when mice were treated with a neurotransmitter, mimicking a nerve signal triggered by stress, immune cells in tissues including the skin and liver stopped moving. How this happens is unknown – behind every 'how' question is another 'how' – but it means there are many ways in which stress weakens the ability of immune cells to react. Nevertheless, stress does not always simply weaken immune responses, as we will see now with allergies.

Does being stressed make common health conditions worse?

Sometimes we belittle allergies. Cancer, heart disease and other illnesses are taken more seriously, probably because allergies are not a top-ten cause of death. Many people with mild allergic asthma, for example, no longer need to think of themselves as having an illness as such; inhalers just become part of their everyday life. But even mild allergies can disrupt our lives. What's more, the sheer number of people with an allergy is staggering – something like 2 billion of us, which is incomprehensible, really – and the number diagnosed is rising year on year.

Allergies occur when the immune system mistakenly reacts against something that is not dangerous, like a protein molecule from nuts or house dust mites. Crucially for our purposes, allergies worsen with stress. In fact, sometimes symptoms only appear with stress.

One extreme situation where this happens – albeit one that affects very few people – is space travel. Many astronauts on the International Space Station have reported a skin rash. For at least one whose blood was analysed, the rash coincided with changes in levels of specific proteins in their blood characteristic of an allergy. Symptoms worsened with doing something especially stressful, such as a spacewalk. In fact, anti-histamines, used to treat allergies, are the second most taken medicine in space after sleeping pills.[23]

Back down to earth, and other studies, looking at many more people, also show a link between stress and allergies. An analysis of over 1,700 people in Germany found that those with anxiety were more prone to seasonal allergies.[24] (Stressed mice are also more prone to allergy, so again this is fundamentally true across life.)[25] On the face of it this is confusing, though, because it counters what we've seen so far about how stress effects immune responses, which is that stress should quieten the immune system down.

So how does an allergy arise? An allergic reaction very often begins with a particular type of human antibody called immunoglobulin E or IgE locking onto the cause of the allergy, such as a protein molecule from a peanut or house dust mite, which in turn triggers the release of the chemical histamine from immune cells. Histamine has many effects in the human body. In a food allergy, histamine can trigger vomiting or diarrhoea, presumed to be the body's attempt to rid itself of the perceived danger. In hay fever, histamine leads to a runny nose and itchy eyes. In allergic asthma, histamine constricts the lung muscles, narrowing bronchiole tubes, making it harder to breathe. So dialling down this unwanted immune response should help. But stress does not help. It only seems to make allergies worse.

There is some evidence that stress hormones stimulate immune cells called mast cells, which are responsible for a lot of histamine's production.[26] In effect this means that stress quietens some of the immune system, but other parts may be sensitised or amplified. In other words, the immune system, just like the financial system, the solar system and all other big systems, is not really one thing, and different bits respond differently.

Autoimmune disease is another condition arising from an unwanted immune response. It is difficult to pinpoint the causes of autoimmune disease, because very often by the time symptoms have developed, immune cells have already been attacking healthy cells for months if not years, making it hard to identify the initial trigger. Genes play a role, but other factors are important too, and stress might be one.[27] Anecdotally, some patients report a stressful event occurring prior to the onset of an autoimmune disease. A survey of over 3,000 people concluded that, although women are more prone to developing autoimmune disease, stress has a greater impact on its development in men.[28] We don't understand why, although in general men and women tend to react differently to stress, psychologically and biologically.[29] Partly this is due to sex hormones, which are thought to be responsible for women typically having more of a tend-and-befriend response to stress, while men, influenced by testosterone, tend to have a stronger fight-or-flight response. All we can say in terms of what happens to our immune health is that somehow stress-triggered hormones may interfere with a person's immune system in a way that occasionally contributes to the development of an autoimmune disease. The effect may be direct, but could be via something else, like a change to a person's microbiome.[30]

The complexity in the effect of stress is simply because the immune system is complex. It must mount an appropriate response to different kinds of dangers, each of which poses its own problems, and germs may reside in different locations in the body. Each part of the immune system must also have its own control measures in place to prevent a reaction against something that is not dangerous. Perhaps it should not be surprising, then, that stress has a multiplicity of consequences for immune health.

Are there specific medical outcomes we can opt for?

Cortisol itself does not worsen allergies, as far as we know, at least not directly. In fact, the opposite is the case: medicines based on cortisol are used to treat allergies. Synthetic versions of cortisol are the active ingredient in preventer inhalers for asthma. Hydrocortisone cream is applied to the skin to treat minor irritations or allergic reactions. A chemical called dexamethasone is very similar to cortisol and about 40 times more powerful in suppressing immune responses.[31] It's used widely, for example to treat rheumatic inflammation, and in 2020, dexamethasone was found to help with Covid-19, making it the world's first officially proven-to-work treatment in the pandemic.[32] For people affected by Covid-19 so badly that they need to be put on a ventilator, dexamethasone reduces their risk of dying by about 30 per cent. Just this use of dexamethasone alone has saved over a million lives. Dexamethasone is also used to treat severe allergies, which only shows how weird and wonderful science and medicine can be: stress can make allergies worse, but a drug based on a stress hormone helps.

Even in simple lab dish experiments complications arise. Most experiments testing how cortisol affects immune cells show that it dampens their activity, but sometimes it seems to have the opposite effect. One example from my own research team has involved studying the effect of the cortisol-like drug dexamethasone on NK cells. When it was in the presence of other proteins commonly found during an immune response, we found this drug reduced immune cell activity exactly as expected. But we noticed something else too. The cortisol-like drug also helped immune cells live longer and react better – not worse – upon repeat stimulation.[33] Experiments in a lab dish are by nature artificial, and it is hard to know how this really plays out in the body, but this result is not easily explained. So we still have a long way to go before we fully understand how even one stress hormone affects the human body. Indeed, cortisol affects the activity of one in five of all 20,000 human genes.[34]

Overall, it seems that stress dampens an immune response when it would be useful, such as against an infection, while at the same time boosting unwanted immune responses such as in allergies. Whichever way you look at it, therefore, excessive stress is bad for immune health. There is every chance too that one feeds into other, because having an allergy or infection is in itself stressful, which may very well make things worse.

Do activities like t'ai chi which help us de-stress boost immune health?

With plenty of evidence that stress affects our immune system, it follows that activities undertaken to help unwind and de-stress should assist our immune health. But here's the big kicker. There's not a lot of clear evidence that things which reduce stress boost immune health. It's not that they don't, just that things which reduce stress are not actually *proven* to help with immune health, simply because it's very hard to test.

For one thing, a fair test is not easy to design: you might consider comparing the frequency with which people get a cold, between those who regularly perform some activity thought to help us relax – t'ai chi, yoga, mindfulness, for example – and those who don't. The problem is there's no way of knowing if any effects are owed to the activity itself, or some shared characteristic in the kind of people who take up something like t'ai chi. A fairer test would be to enrol people and split them up randomly into two groups, one given a relaxation technique to practise, the other not. Ideally the control group – those not set the activity in question – would be given something else to do to replicate other possible benefits, such as time spent with a social group. But this couldn't easily be set up as a 'blind' trial in which nobody knew whether they were taking the 'medicine' being tested or not: people would know whether they were performing t'ai chi or not, so the placebo effect – your brain convincing your body you are doing something to help – could influence the outcome.

A second issue is money. Studies assessing what happens to people performing something like t'ai chi usually involve small numbers, something like 50 or so individuals.[35] By contrast, it's not uncommon for thousands of people to be enrolled to test a new pharmaceutical drug, which must also be compared to other existing medicines – one reason why the average cost of bringing a new drug to market is something like $1 billion.[36] There simply isn't that kind of budget available for testing the effects of mindfulness or yoga on immune health.

A third problem, and probably the most difficult to address, is how to determine whether relaxation affects immune health. A common measure is the amount of a specific antibody present in a person's blood after they've been given a vaccine. But what this means in terms of how well a person will fight off an actual infection is hard to know. Antibodies are important to how we fight an infection, but as we've already mentioned, it's not obvious how a 10, 20 or 50 per cent increase in antibody levels will impact a person's well-being when infected.

Indeed, there isn't really *any* single number from a blood test that can reflect the status of a person's immune health. First, because the immune system is so multi-layered and complex.[37] Secondly, because the immune system in a person's tissues or organs, say their nose or lungs, is also vital. The state of immune health within a person's organs or tissues is only somewhat reflected by the cells and molecules circulating in the blood. So you can see why designing a trial to test our level of resistance to an infection is not easy, without the actual infection being given deliberately to see how we fare – which isn't ethical, at least on a scale large enough to test the effect of relaxation methods.

On the positive side, various studies show that all sorts of things associated with relaxation, from t'ai chi to reading, do reduce a person's cortisol level, which during long-term stress should, from everything we've discussed, be beneficial for immune health. Mindfulness lowers cortisol.[38] Even a 20-minute nature walk – defined as being anywhere outdoors where participants felt they were interacting with nature – lowers cortisol.[39] Hence the idea of prescribing a 'nature pill'. A positive attitude also helps: if people have a positive mindset about stress before doing a stressful activity like public speaking, including understanding how it could enhance their performance, cortisol levels spike less.

More broadly, people who instinctively tend to be

optimistic spike less cortisol in stressful situations. One analysis of nearly 8,000 cortisol samples taken from 135 over-60 year-olds across six years found that those who tended to have an expectation that things will turn out well, as determined by a questionnaire, were less likely to exhibit higher levels of cortisol on days they felt stressed.[40] Another way of assessing the power of positive thinking is to use the so-called *Best Possible Self* intervention. For this, participants answer a worksheet about themselves describing their best possible future, to improve mood and increase optimism. Doing this causes participants to spike less cortisol when given a stressful task.[41] Perhaps in the future cortisol levels will be a useful thing to measure frequently, like our pulse, blood pressure or steps walked, to check on our well-being.

So what should I do about stress?

Understanding the science of stress and immune health surely helps us make informed decisions, especially because business has a vested interest in how we deal with stress, making money from helping us de-stress and so on. Which brings us to a rather shocking plot twist in Hans Selye's story.[42] Never disclosed in any of his books or public talks, but only revealed many years after he died, when paperwork was scrutinised, is that Selye was in cahoots with the tobacco industry.[43]

Selye's ideas were useful to the tobacco industry because cigarettes could be sold as stress relief. He received significant funding for his research and in return, over the radio, spoke about the benefits of smoking for people under stress. The tobacco industry also used Selye's ideas to suggest that if any increase in cancer or heart disease risk was attributable to smoking, it could be down to smokers having busy and important lives.

In 1959, Selye was paid to write a memo that was used in legal proceedings, in which he stated that a correlation between cancer and smoking does not prove that smoking causes cancer. Though strictly speaking correct, it was misleading: it had already been established that chemicals in cigarette smoke cause cancer in animals.[44]

The idea that smoking may help reduce stress was used by the tobacco industry long after Selye died, and again something of this has persisted. A lesson for us all is that we must be very careful about vested interests – not always easy. For me there's another lesson: that there's always more to the story – in this case the life of the researching scientist himself – than first appears.

The bottom line, however, is that long-term excess stress unquestionably leads to a dysregulation of the immune system, which causes problems, lowering our responses to vaccination and increasing the symptoms of allergies. Some of this is down to increased levels of the stress hormone cortisol. Various practices, from t'ai chi to walking amid nature, can reduce cortisol levels, but proving they help immune health, though it makes sense that they would, is difficult.

The point at which stress becomes a problem is hard to pin down. It's like asking how a goal happened in a game of football. What precisely caused the ball to go over the line? Certainly the person who last kicked the ball. And the person who kicked the ball before that. And the fact that the goalkeeper dived too late. But there's also the selection of the team, the tactics, the manager's speech, the training sessions, the noise of the crowd, the way a physio dealt with a player's injury a couple of days ago, and a conversation on the phone which led to a new striker being bought six months earlier. The point at which any of these causes led to an effect is elusive and vague. And human biology is so much more complicated than a football match.

There is this kind of depth and uncertainty across much of human biology. Like love or happiness, stress is multi-faceted and hard to measure. Still, from what we do know, it seems a very good idea for our immune health to, at least sometimes, with good conscience, rest.

6

Inner Beauty Sleep – or, does sleep help immune health?

Sight, hearing, smell, taste and touch are the most obvious inputs to our brain, helping us respond to whatever situation we find ourselves in. But like a secret sense, our immune system provides an input too, albeit less up-front and more subtly, informing the brain about danger or damage within us. All kinds of immune cells produce protein molecules which influence the brain, from which we can 'feel' our immune system getting to work. Often we feel sleepy when we're ill – why? Is sleep perhaps triggered in order to help our immune system work well? And does this mean that in general, sleep duration affects our immune health? These are the important questions we will turn to now, beginning with exactly *how* the immune system communicates with the brain, then *why*, and finally what this means for our lives.

How does the immune system make us sleepy – and why?

How does any part of the human body tell another part what's happening? With protein molecules which circulate in the blood. The vital protein molecules which link our immune system to our brain are called cytokines, and they function as the immune system's hormones.[1] You may not have heard of them before – antibodies and hormones have entered the lexicon of popular culture, but not cytokines – yet they are just as crucial to our health. All immune cells produce cytokines, and to a lesser extent other cells too. There are well over a hundred different kinds. Collectively one of the most important things they do is to shape an immune response according to the type of problem manifesting in the body. For example, if a viral or a bacterial infection is underway then cytokines will circulate in the blood to switch on immune cells that are good at dealing with viruses or bacteria. If there's a worm infection, different cytokines are produced, mobilizing a different set of immune cells.

Messages relayed by cytokines can be complicated. Some cytokines can dampen, enhance or completely change the actions of others. Networks and feedback loops amplify the right type of immune response, which may need to change over time, and cytokines trigger appropriate repair and healing processes too. Amazingly, some germs produce their own versions of our cytokines, to send the human immune response off in the wrong direction.[2] Tumours produce cytokines too (and other kinds of hormones), which can also confuse the immune system, or just stop it altogether.[3]

Cytokines are also the target of several medicines. For example, one of the main blockbuster drugs for rheumatoid arthritis, called anti-TNF therapy, works by blocking the action of a particular cytokine called TNF.[4] Blocking this cytokine cuts off a line of communication, stopping immune cells from egging one another on, decreasing joint inflammation.[5] Crucial to our purposes here, neurons in the hypothalamus, hippocampus and the brainstem – areas of the brain involved in regulating sleep – also directly sense cytokines.

From this it follows that particular cytokines promote the first stages of sleep, and not only when we're ill. Throughout the day, blood cytokine levels oscillate, bringing the body through all manner of peaks and troughs. Working out exactly what goes up and down, and when, has turned out to be difficult because people vary so much, but there's a rough pattern all the same.[6] For example, one of the cytokines involved in regulating sleep (called IL-6), tends to dip just as it's time to wake up.[7] During an infection, however, certain cytokines rise beyond their everyday flux, and this is what makes us feel more tired than usual.

Cytokines don't just happen to rise and fall concurrently with sleep, but affect it directly. We know this because injections of cytokines alter sleep patterns in animals.[8] For instance, an injection of them has been shown to increase the duration of the quiet and restful phase of sleep in rats, mice, monkeys, cats, rabbits and sheep.[9] We also know that neurons in the human brain have receptor protein molecules shaped to lock onto cytokines, and when they do, this triggers signals to emanate out from that neuron to other neurons, beginning a cascade of synaptic communication.[10] In turn, certain cytokines increase a period of sleep in us called slow-wave sleep, which is what we commonly think of as 'deep sleep', the stage of sleep when the body is most immobile, everything slows down and memory consolidates.[11] (Which means that, feasibly at least, the actions of your immune system could even affect things as profound as what you remember.)[12]

It isn't only the presence of an infection that increases blood cytokine levels. Getting too little sleep tends to increase them too, which is one way by which a lack of sleep leads to a desire for more sleep. Clinical problems with sleep elevate cytokines too, such as narcolepsy, a rare brain condition which can prevent a person from choosing when they sleep.

Interestingly, several studies have found that experiencing a tragedy, or a social threat like bullying, or feeling lonely, also tends to increase the levels of certain cytokines in the blood, again making us feel sleepy.[13] All this suggests that the body's response to many different kinds of problems is to elevate cytokine levels, which encourages sleep. So it seems sleep is part of our self-defence against all sorts of trouble.

Exactly why is hard to answer, because essentially why we sleep in the first place is still mysterious. It is one of the most fascinating questions of contemporary biology to which we don't yet have an answer.[14] (One long-standing theory is that sleep gives time for the brain to clear out toxins, but recently this has been thrown into question by an experiment suggesting that a dye was cleared more efficiently from the brains of mice while they were awake.)[15] But while sleep remains mysterious, we do know a lot about why we have an immune system: including to fight germs and help heal the damage caused by germs. And importantly, the stress hormone cortisol is kept low while we sleep. Since cortisol dampens immune activity, as we discussed in Chapter 5, then by keeping stress low, sleep helps our immune system. This leads to the inference that our immune system is more powerful while we sleep.

Indeed, the total number of immune cells in blood does tend to be boosted during the night. It is possible that this happens because our immune system has adapted to be different during the day and night, to deal with specific night-time threats. For example, many types of mosquito are more active at night. Then again, other types of mosquito, including those that transmit Dengue and Zika viruses, are active during the day, so overall this seems unlikely to be true. We don't really know why the immune system changes at night. It might be because so many processes in the human body change at night — we stay relatively still, body temperature decreases, hormone levels change, and so on — that the immune system is just inevitably affected. (There's an important general point here: lots of things about the human body do not operate to an optimal design, but just happen to be that way on account of something else.)[16] There are nuances too, in exactly what happens to our immune system while we sleep. Some types of immune cell, such as certain kinds of T cell, are more abundant in the blood during the daytime.[17] The whole idea that the immune system is more powerful at night is, therefore, too simple; it is more accurate to think of our immune system as being in a *different state* while we're awake compared with when we're asleep.

What is the right amount of sleep?

Is there an optimal sleep duration? If we get too little sleep, our ability to concentrate is clearly affected,[18] and this is much easier to measure than immune health. In one experiment, nearly half a million adults were asked how much sleep they tended to get, and then play online games to test their powers of cognition. The games included memorizing the position of six matching pairs of cards to test the participants' memory, and a version of the game Snap to assess reaction times. How well they did in these tasks correlated with the amount of sleep people had had, with the optimal duration being seven hours. Less sleep correlated with worse test scores. More surprisingly, too much sleep also correlated with worse scores.[19] The results followed a bell-shaped curve: performance decreased every hour above and below 7.

Before we all start making sure we get exactly seven hours' sleep, no more, no less, there's something to bear in mind. Seven hours of sleep looks optimal when everyone is plotted on the same chart, but it doesn't mean that seven hours' sleep is optimal for you individually. Needs vary.

We're all aware that sleep varies dramatically between people, including that some of us prefer to go to bed early and wake up early, while others like to stay up late and get up late. In scientific jargon, this is our chronotype.[20] When looked at carefully, however, it's not that people fall into two types, larks and owls, waking up early or late, just that there's a continuum of wake-up times different people naturally prefer. Interestingly, many gene variants which are more common to morning people can be traced back to the Neanderthals, a now-extinct species of human who geographically overlapped with modern humans for over 30,000 years when modern humans migrated out of Africa. During this time Neanderthals and humans interbred, and hence there are many Neanderthal-derived genes present in people today, especially people outside Africa,[21] some of which probably helped humans adapt to their new environment.[22] For example, as days became shorter there was probably an advantage to being awake early, to start collecting food.[23] This might appear by-the-by, but it's also a window into who we really are – and an insight into where our sleep preferences come from.

Many other things affect our sleep preferences too, age especially.[24] Babies, children and teenagers sleep more than adults. People aged 20 tend to stay up around two hours later than people in their mid-50s (my kids' age and my age, so I know it's true; they do like to phone me at night just as I'm falling asleep).[25] Other factors include the jobs we have, the times at which it's daylight, and any number of social pressures that make us stay up later or get up earlier than our natural tendency. Sometimes you just want to sleep but can't stop thinking. Hormones including oestrogen and progesterone also affect sleep, which is one reason women often feel tired in early pregnancy (with other factors coming into play later in pregnancy too). About 9 out of 10 adults in the US have a caffeine-laced drink almost every day. A cycle can build up in which caffeine can improve certain types of performance, but also disrupt sleep, thereby worsening performance, and leading to more caffeine taken to aid performance, and so on.[26]

Magnetic resonance imaging (MRI) has been used to look for any differences in people's brains according to how much sleep they have. Several studies have found that the amount of sleep a person tends to get correlates with how large particular regions of the brain are.[27] This isn't quite as striking as it sounds, though, because various research teams have arrived at different conclusions as to exactly which parts of the brain vary with sleep. Even if these can be reconciled, it would still be hard to know if sleep influences brain structure, or if people with certain brain structures tend to sleep more or less than average. There may be no causal link at all: something else may affect both brain structure and sleep. We do know that sleep deprivation affects synaptic connections in the brain, some weakening and others strengthening, resulting in a loss of attentiveness, memory, judgement and decision-making.[28] But like many things about the brain, it's only slightly understood.[29]

Another thing commonly experienced, and more pertinent to our discussion of immune health, is that sleep affects the symptoms of illnesses. Most often this comes down to fluctuating levels of hormones or cytokines. For example, problems from asthma are more common at night, and deaths due to asthma are more likely around 4 a.m.[30] Indeed, unwanted immune responses tend generally to worsen at night, at least in part because the hormone cortisol is kept low while we sleep, which means immune responses are not suppressed by it.

This is important for rheumatoid arthritis, where the immune system attacks the tissue lining of joints, causing unwanted inflammation and pain. Again, because cortisol is kept low, inflammation can build up while we sleep, leading to stiffer joints in the morning. In fact, symptoms being worse in the morning can be used to diagnose an active inflammation causing the problem in the first place,[31] which is important because it means that anti-inflammatory medicines, or medicines which block a cytokine like anti-TNF, may help.

There's a paradox here, though. If sleep makes symptoms from unwanted immune responses worse, it ought to follow that having less sleep would help. But it doesn't. Disrupted sleep is especially problematic – for people with sleep apnoea, for example, in which breathing becomes shallow, or stops and starts. A short-term loss of oxygen, which can be caused by sleep apnoea, raises several cytokines in blood. Essentially this is because low oxygen is sensed in the body as a danger and, as with other types of danger, the body responds by elevating certain cytokines.[32] These switch on various processes in the body, which can include a heightened level of activation of the immune system. This fits with people experiencing disrupted sleep having worse symptoms from autoimmune diseases, including lupus, rheumatoid arthritis, ankylosing spondylitis and psoriasis.[33] This is no trivial matter: almost two-thirds of people with rheumatoid arthritis also have difficulty sleeping.[34]

One extreme experiment with animals makes the point dramatically. Mice were made to stay awake for several days by placing them in a cage ankle-deep in water, which prevented them from assuming their normal sleep position. After four days, 80 per cent of them died.[35] Their immune system had gone so dramatically awry, and cytokine levels climbed so high, that the mice experienced intense inflammation all over, and their organs were severely damaged.

Surprisingly, for people staying up all night – just one night – their blood immune cells start to exhibit some of the tell-tale signs of being involved in autoimmunity.[36] Whether this has any real lasting impact on our health is not clear, and it's certainly not the case that one night of staying awake triggered any actual autoimmunity in anyone. We also don't know whether the following night's sleep quickly brings the immune system back to its normal state. Interesting as this is, then, we need other ways of studying the effects of sleep on immune health. One useful way of trying to assess the effects of staying up, and how much sleep is needed, is to look at people working night shifts.

Shift workers often sleep during the day rather than at night, but also tend to sleep less overall: daytime sleep after a night shift tends to be short, around four to six hours.[37] This is very likely to affect immune health.[38] One study found that shift workers had a small increased risk of multiple sclerosis,[39] which was more pronounced among people working at night from a young age, perhaps indicating that health consequences build up over time. Another analysis shows a slightly increased risk of breast cancer.[40] While these findings seem to suggest it's best to sleep at night, for a good stretch of time, it's difficult to make any generalisations, as shift workers aren't really just one group: they work in all sorts of specific environments, from offshore oil rigs to hospitals, and may well have other personal differences such as diet, levels of exercise, stress and so on. But on balance there is evidence that disrupted sleep, and too little or too much overall, can be problematic for immune health.

Can sleep reduce our risk of cancer?

In an experiment performed over twenty-five years ago, now considered a landmark study, immune cells were isolated from blood samples taken from volunteers who were deliberately kept awake from 3 to 7 a.m. Then, in a lab dish, where things can be controlled and examined precisely, these isolated immune cells were tested for how well they could kill cancer cells. Compared to immune cells taken from people left to sleep normally, they were much less efficient.[41] After a night of normal sleep, the volunteers' blood immune cells returned to normal.[42] On the face of it this seems to imply that sleep is extremely important for our immune system to able to fight cancer.

Experiments in a lab dish, however, are far removed from the full complexity of the human body, or any living organism for that matter, and so have many limitations. One – especially for considering our immune system's ability to fight cancer – is that by mixing immune cells with cancer cells in a lab dish, the issue of whether immune cells can reach or penetrate a tumour has been circumvented. In mice, long-term sleep deprivation not only reduced the total number of immune cells able to kill cancer cells, but also decreased the numbers of immune cells able to reach and penetrate a tumour in the first place.[43] So whatever happens to their ability to kill, it wouldn't matter so much if they are not in the right place to attack.[44] From this, it seems, there are many ways in which sleep disruption can affect immune health in the context of cancer.

But like a novel where things twist and turn, page after page, and you eventually start to anticipate another side to the narrative, this is not the end of the matter, and this is where things get controversial. In this case, the plot twist comes from trying to determine whether sleep affects the risk of cancer overall. Put another way, across large numbers of people, are sleep duration and cancer risk correlated? The answer is, different studies have come to different conclusions. In fact, any of the possible outcomes can be found in one research report or another – that a short or long sleep time correlates with lower or higher risk of cancer, or that it makes no difference. Which means that for a headline or soundbite you could cherry-pick the data and say what you like.

A systematic review, collecting the data from 65 different trials meeting certain criteria and analysing them together – amounting to an analysis of more than 1.5 million people – set out to determine once and for all the effect of sleep time on cancer.[45] The overarching conclusion turned out to be straightforward: ultimately, neither short nor long sleep duration changes the overall risk of cancer developing.

However, two subtleties can be picked out. First, that long sleep duration correlates with a slightly increased risk of one type of cancer, colorectal cancer. The effect is small and could be due to any number of things, including the possibility that sleep times may be affected by other diseases or undiagnosed conditions, or relate to a person's socioeconomic status, or less physical activity, any of which may also correlate with increased risk of colorectal cancer. The second nuance came from breaking up the analysis according to where people lived, from which it emerged that short sleep times particularly increase the risk of cancer in the Asian population.

Caution is needed, however. Again the effect was small, and different methods might have been used according to where the study was done. Still, it raises the possibility that different groups of people could be affected differently by sleep times. This might relate to genetic variation, culture or the environment – for example, some people tend to nap frequently, eat or drink in bed, sleep with the light on or off, while it's noisy or quiet, and so on, all of which will vary according to where you live. More research is needed to check these subtleties and, if anything holds up, understand them better. There is plenty of room for new advice and guidance to emerge as the research continues.

There's also the question of whether sleep affects a person's well-being, or chance of survival, once diagnosed with cancer. This has been looked at far less than the risk of cancer arising in the first place, and here again there are many things we don't know. Might sleep help people being treated with immune therapies? This seems feasible, and would fit with the experiments in a lab dish, and in animals, showing that disrupted sleep weakens immune responses against cancer, but hasn't been looked at specifically.

Even if sleep can help, that doesn't mean it's easy to *get* a good night's sleep in the face of cancer. In fact, sleep problems are especially common for people with cancer, with evidence that the presence of cancer itself may disrupt the brain circuitry which regulates sleep.[46] This could arise from cancer cells themselves producing cytokines, which act on the brain to interfere with sleep patterns, making it harder to sleep normally. (I'm just speculating here, but this might be advantageous to the cancer because of the detrimental effect disrupted sleep has on the immune system.) Then there's the stress of simply having cancer, which takes its toll too, affecting the immune system directly, and affecting sleep.

Many cancer treatments, from radiation to chemotherapy to surgery, can affect sleep as well. Cancer patients often feel fatigue as a side-effect from treatments, perhaps owing to changes in cytokine levels, or by losing appetite, which can decrease energy levels.[47] Even cancer survivors are more likely to suffer problems with sleep compared to the general population.[48] All this remains underexplored.[49] As anyone who's been unfortunate enough to experience cancer knows, or if you've witnessed cancer take hold of a loved one, it's certainly not easy to get a good night's sleep.

Does a good night's sleep help with the effectiveness of vaccines?

The effect of sleep on our response to vaccination is more easily measured, and thus easier to draw conclusions about health from. It was in 2002 that a landmark experiment was carried out to test this. Volunteers were divided into two groups, one to be sleep-deprived, the other not. Specifically, the first group were only allowed to sleep from 1 to 5 a.m., for six nights. Following that, they were allowed to recover by sleeping twelve hours a night for the next seven nights. In the morning after their fourth night of short sleep, they were given a shot of a flu vaccine. The other group received the exact same vaccine while sticking with their usual sleep times. The effect of being sleep-deprived was striking. Ten days after being vaccinated, members of the sleep-deprived group showed blood antibody levels less than half measured in those who had been sleeping normally.[50] Quite simply, disrupted sleep made the vaccine work not even half as well.

Since this first study, many similar experiments have been reported, and an overall picture is clear. Sleeping less than six hours a night leads to lower levels of antibody being produced by a vaccine.[51] This doesn't mean that poor sleep stops a vaccine from working entirely: it just lowers the amount of antibody produced. It's not easy to know what this means in terms of the level of protection a vaccine gives, as it's hard to translate a specific rise or fall in antibody level into a specific gain or loss in protection against a real infection. Producing half the amount of antibody as somebody else could make you much less protected, or may only have a small effect, because the chance of an infection, and the severity of symptoms from an infection, vary as a result of so many factors.

Interestingly, the effect of sleep on vaccine responses isn't found to fluctuate at all if people are simply asked how much sleep they have. That's because people get it wrong when asked about their own sleep time, which muddles the results. It's hard to know, for example, how much time is spent lying in bed but not asleep. A correlation with vaccine responses is only evident when sleep is monitored precisely, in a lab setting or by using a sleep-tracking wearable device like a watch or ring.[52] Even more interestingly, the impact of sleep on vaccine responses is much bigger in men than women. It's not obvious why, but women tend to respond better to vaccines anyway – one study found that women given a half dose of flu vaccine responded to the same extent as men given a full dose.[53] Why women differ from men in their vaccine response in general, or why sleep may disproportionately affect vaccine responses for men, remains at the edge of knowledge, with every discovery raising new questions.

Antibody levels triggered by a vaccine naturally drop over time, so one way of understanding the impact of poor sleep is to consider the equivalent time it would take for them to fall by the same amount anyway in the general population. Such analysis shows that poor sleep reduces the effectiveness of a Covid-19 vaccine by the same degree it would naturally decline in about two months.[54] So it has an effect, but not a pronounced one, and certainly not to the extent that we should not take a vaccine *unless* we're sleeping well. Overall, it's hard to know how big an effect sleep really has, and it might be quite small, but on balance, around the time of a vaccination, it probably helps to some extent to try and get a few good nights' sleep.

What can I do?

In summary, we have three ways of testing the effects of sleep on immune health. First, people are deprived of sleep, and we then isolate immune cells from their blood and test how well they react, for example, to cancer cells. Or we give them a vaccine, and measure their antibody levels. Secondly, animals can be made to stay awake for a prolonged period, and we assess what happens when they are then exposed to bacteria or a virus or injected with cancer. Thirdly, we can analyse populations of people to see whether health issues correlate with variations in when and how people sleep.

Each approach has its own caveats and difficulties, which is important to bear in mind whenever you come across new advice about sleep and immune health – like the idea that we should aim for seven hours' sleep, based on it being optimal for people doing well at games which test their memory and cognitive skills. It might look optimal when half a million people are plotted on a graph, however, but might not be for you personally. Other issues include the fact that people are not very good at stating accurately how much sleep they get, that some people are larks and others owls, and what's the consequence of some people napping in the daytime? Neither is it easy to monitor different stages of sleep, rather than just a person's total sleep time. What emerges is that the precise answer to something as simple sounding as 'How much sleep should I have?' is not straightforward at all.

A compounding factor is that sleep is entwined with so much else. The elderly often struggle with sleep, for example. Stress, too, influences sleep. Age and stress both have their own impact on the immune system. It's almost impossible to isolate the effect of sleep, and sleep alone, from everything else connected to it. And yet the matter couldn't be more important. As many as one in three adults feels overly sleepy in the daytime,[55] and one in four people in England says they suffer from insomnia.[56]

Which brings us to the question of *how* to sleep well. For some people a sleep clinic can help. If we return to sleep apnoea, treatments can include a machine which delivers pressurised air through a mask to keep the airways open. For more common, minor difficulties, any number of books, as well as countless magazine articles, give advice, which amounts to: sleep in a place that is quiet, dark and relaxing. Avoid caffeine late in the day. Don't keep your mobile phone by your bed. Don't have a large meal too late in the evening. Be physically active during the day. Be consistent: go to bed at the same time every night and wake up at the same time, training your body into a routine.

One recent study looked at whether odours can improve people's sleep.[57] Three were tested: orange, a sweet, floral scent and a complex commercial perfume (citrus, floral, musky). On average none of these odours had any impact on sleep quality. But if a person rated a particular odour being especially pleasant, they tended to perceive their own sleep quality as improved by it. In other words, none of these odours had any special ingredient, like a drug, to make us sleep well, but some might help, according to what *you* sense as pleasant. Of course, understanding what we smell as pleasant is yet another realm of science, and unquestionably derived from experiences, remembrances and imaginings. All deeply personal.

Which brings us to one final point about sleep. A more personalised analysis of sleep is where we're heading. Sleep-monitoring technologies are advancing rapidly. As yet, few are robustly validated and readily available or affordable, but this looks set to change, because companies are investing in new tools for sleep assessment, from non-contact technologies embedded within mattresses to wearable watches, armbands and in-ear devices. As our understanding improves, and sleep monitoring becomes more precise and available, all sorts of everyday metrics will become easily available at home: sleep times, different sleep stages, body movements, breathing, snoring, heart and respiration rates, body and room temperature and more.[58] Then it will become, if it isn't already, a matter of personal choice as to how much you want to know. Personalised sleep analysis could become enabling and powerful – a tool to help improve your immune health – or just something else to worry about.

SELF DEFENCE

Like a wave building up, far out in the distant ocean, our conception of sleep health is just beginning to take shape. We vary in our needs. We vary in what helps us sleep well. For now we can leave it at: whatever works for *you*, to help *you* sleep well, will also help your immune health.

7

The Two Big Systems – or, does immune health affect mental health?

The seventeenth-century French philosopher Descartes considered the mind and body to be such different things that their interaction was impossible. Obviously, we now know this is not true. Specific to our purposes here, immune health and mental health are very much linked. One thing that has helped us realise this is seeing what happens to patients' moods when they are taking a particular medicine for autoimmune disease.

In the early 1980s, the Polish-born scientist Marc Feldmann wondered whether some autoimmune diseases might be caused, at least in part, by immune cells activating one another through their cytokine secretions to such an extent that that activation becomes a self-perpetuating vicious circle, so that inflammation persists and the body's healthy tissues are damaged.[1] In London, Feldmann met the Indian-born clinician Ravinder Maini, and together they focused on rheumatoid arthritis, which at the time was considered such a complicated disease that no simple treatment would ever be likely to help.[2] In 1992, they conducted a small trial at Charing Cross Hospital in London, to see whether a medicine blocking the actions of a cytokine could help alleviate rheumatoid arthritis.[3] In talks reminiscing about what happened in that trial, Feldmann often shows a video of patient number eight.[4] Before the treatment, she walks up and down stairs holding onto the handrail, moving very slowly, clearly in pain. After treatment she runs down the same steps as fast as anyone could with an exclamation of 'Ta-da!'

As more and more people were given this medicine, many said it made them feel better mentally, very quickly. As one patient put it, 'Here I was up and away and feeling so much more confident in myself.'[5] Nurses wanted to set up an infusion for patients because they knew it would make them feel happier and full of gratitude.[6] This feeling seemed to come more quickly than could really be expected from the time it took for their joints to improve, raising the possibility that blocking this cytokine was improving their mental state as well as helping them physically improve their immune condition.

Feldmann and Maini couldn't have foreseen just how important this medicine would become. It is called anti-TNF (TNF being the cytokine it blocks), and nowadays is used widely to help stop unwanted inflammation in all sorts of situations, not just rheumatoid arthritis. Crohn's disease or colitis, for example, where there is inflammation in the digestive system, or psoriasis, where there's inflammation in the skin.[7] While it is not directly prescribed for any mental health condition, it is one of the world's most prescribed medicines, which gives us the chance to look at whether it helps a mental health condition people happen to have alongside the physical condition they have been given the medicine for.[8]

It turns out that blocking this cytokine can indeed help alleviate fatigue and depression in people with rheumatoid arthritis,[9] and also helps with anxiety or depression for patients with inflammatory bowel disease. This fits with the suspicion that blocking this cytokine may help with at least some people's mental health condition. Direct evidence for this comes from MRI scans showing changes in brain activity within twenty-four hours of taking the medicine.[10] At this stage, so soon after treatment, swelling of the joints and other clinical markers of inflammation show no alteration.

These observations have caused us to think deeply about whether immune activity might sometimes cause or exacerbate depression and other mental health conditions.[11] We could be on the cusp of a revolutionary new approach to understanding and treating some people's mental health problems. But first we must carefully assess just how strong the connection between immune health and mental health really is. How exactly are they related?

How is immune health linked with depression?

It is well established that, for people with illnesses that involve activation of the immune system, such as allergies and autoimmune diseases, there is also an increased risk of depression.[12] For example, the likelihood of depression developing in people with diabetes is about double the normal rate.[13] For patients with rheumatoid arthritis there's a 75 per cent increased likelihood they also suffer from depression compared to the general population.[14] For those with multiple sclerosis, around one in five also has depression.[15] These correlations could come about because the physical symptoms affect a person's mental health, or because the diagnosis itself is psychologically challenging, especially for a condition which is incurable, with symptoms only going to worsen over time. But it's also what we would expect to see if depression, and other mental health issues, can arise directly from inflammation.

In medical practice, depression is noted differently if it occurs in the context of another medical issue that could explain it. In this case, a patient is diagnosed with comorbid depression. In fact, comorbid depression is one of the most common problems alongside all other diseases, not just those that directly relate to immune system. For example, Parkinson's disease and hypertension increase a person's risk of having depression to a similar extent as type 2 diabetes or rheumatoid arthritis.[16] Comorbid depression is usually treated just as any depression would be, except that certain antidepressants are avoided if they might have an adverse effect on the patient's other medical problem.

In general, more and more people are being diagnosed with more than one condition. To some extent this might be because we are increasingly recognising, naming and diagnosing illnesses and conditions, but it also seems likely that some health problems are becoming genuinely more common, especially mental health issues. The current trajectory is that by 2035 two-thirds of older adults will have multiple health issues.[17] Already, by the time we reach seventy-five about half of us will have developed, or experienced at some time in our lives, at least one mental health disorder.[18]

In terms of understanding how much of this is linked to immune health, distinguishing whether one issue causes another, or a person just happens to have several conditions, is not always easy.[19] One reason for this is that the immune system, its activation and any ongoing inflammation, locally or across the whole body, influence so many other conditions and diseases. There are also many traits and lifestyle factors that affect the risk of certain diseases – obesity, for instance, increases a person's risk of depression and anxiety, and the other way around: depression and anxiety can contribute to obesity. All three affect the risk of type 2 diabetes.[20] In fact, it has been predicted that by 2027, 33 per cent of females and 15 per cent of males with type 2 diabetes will also have depression.[21] Everything is so intertwined.

The complexity of how diseases arise, from genetic variations to where and how we live, and the way symptoms affect other symptoms, makes it very hard to put any person's specific medical situation into an orderly linear trajectory. Still, there is evidence that sometimes high cytokine levels in the blood might be partly responsible for multiple problems. For example, people with both depression *and* asthma tend to have a higher level of the cytokine TNF in their blood, compared to having asthma or depression on their own.[22] Similarly, people with multiple sclerosis *and* depression have slightly higher numbers of immune cells producing cytokines in the blood compared to people with multiple sclerosis *but not* depression.[23] These findings fit with cytokine levels being a factor in whether depression sets in alongside these other illnesses. It is interesting too that one cytokine, TNF, seems especially important, despite there being well over a hundred cytokines overall. This cytokine is probably a focal point in a network, or at the top of a cascade, for its level to have a particularly significant effect. Indeed, both in patients and experiments with cells in a lab dish, blocking the cytokine TNF, exactly as in anti-TNF therapy for rheumatoid arthritis, stops the production of other cytokines too.[24]

Looking back through patients' medical history – including when they started taking anti-depressants – reveals that depression or anxiety is more likely than average for people who will later go on to develop an immune-mediated disease such as inflammatory bowel disease, multiple sclerosis or rheumatoid arthritis.[25] This increased risk of a mental health issue is evident as early as five years before the immune condition is diagnosed.[26] This could simply relate to the difficulty in diagnosing these immune conditions but, taken at face value, it fits with the notion that a rise in cytokine levels caused by a developing problem with immune health affects a person's mental health before physical symptoms begin to manifest.[27]

Needless to say, depression is far more widespread than just afflicting people with allergies or autoimmune disease. But here, too, high levels of cytokines in a person's blood correlate with depression arising in the future.[28] For example, women over the age of fifty with no sign of depression but higher-than-average markers of immune activity in their blood tested on two separate occasions showed an increased risk of developing depression four years later.[29] More strikingly, a study of 15,000 children found that those with higher-than-average levels of the cytokine IL-6 in their blood at the age of nine, with no sign of depression, were more likely to suffer depression many years later at eighteen.[30] All this indicates that at least for some people, high levels of cytokines in the blood can prefigure the onset of mental health issues. These timeline data imply that inflammation precedes depression, which is consistent with – although not proof of – one causing the other.

This begs the question as to whether a rise in cytokine levels, which occurs when the immune system fights an infection, sometimes relates to the onset of mental health problems. An analysis of fifty-nine different studies comparing what happens to people after they've had a viral infection showed that anxiety, distress, depression and other mental health problems were substantial, even after the infection itself had seemingly resolved.[31] This includes people recovering from Covid-19, flu, SARS, MERS or Ebola. Again, this supports the notion that an immune response can directly affect mental health.

What's more, cytokines at elevated levels in people with depression include those we have come across earlier for their ability to also regulate sleep (IL-1 and IL-6),[32] which could in part, account for why about three-quarters of people with depression have trouble sleeping.[33] In fact, insomnia and sleep problems are what many people with depression seek specific medical treatment for.

Other evidence for how immune and mental health are linked comes from studying animals. Incredibly, giving rats an injection of the cytokine IL-1 induces both a fever and behaviours reminiscent of depression, such as less movement and less interest in other rats introduced into the cage. IL-6 also induces a fever in rats, but does not trigger the same behaviour changes.[34] Still, things are complicated, because when IL-6 and IL-1 are both given to rats, the effects of IL-1 on behaviour are magnified – so each cytokine doesn't act alone,[35] meaning that assessing someone's mental health purely from an analysis of cytokine levels in their blood is likely to be extremely hard. Still, the basic observation is more important: cytokines can directly elicit depressive behaviours in animals.

Another way of testing for a link between immune and mental health is to use animals that have been genetically altered, to test whether genetic manipulations known to affect immune responses also affect behaviours. In one experiment, mice were placed in a cage with an aggressive strain of mouse for ten minutes every day for ten days.[36] The normal mice became 'depressed', in the specific sense that they subsequently spent more time away from other mice and less time in open space. But, remarkably, mice genetically altered to have a deficient immune system did not develop these stress-induced changes in behaviour. Another study used a different genetic manipulation to create mice which lacked a large part of their immune system, and this led to them being less social with one another irrespective of any stressful situation being invoked.[37] If these mice were given immune cells from normal mice they recovered their ability for social interactions.[38]

Quite astonishingly, normal mice – those without any genetic modification and a normal immune system – could also be stopped from developing anti-social behaviours after stress if they were treated with a medicine to block the cytokine TNF, the same type of medicine used to treat rheumatoid arthritis patients. This certainly fits with the theory that high levels of this cytokine can cause complex behaviour changes. Arguably, then, the bottom line is that animal behaviours which mimic some aspects of a human mental health condition are clearly influenced by the immune system.[39] Mice aren't able to self-reflect or ponder their situation in the same way we are, and so it's impossible to really assess the mental health of a mouse or rat, or even know what such a term means.[40] But the fact that they too are less social, more withdrawn and less exploratory, when experiencing an inflammation, or even when injected with a cytokine, suggests that this happens as a result of some important innate, basic or primal process in animals and us.

In fact, it's possible these connections with behaviour have evolved as part of our defence. For example, if an immune response causes us to be less sociable, this will help prevent the spread of an infection from one person to another. But if this process overshoots, or continues for too long – and I'm only speculating here – it might lead to long-term problems, which we then consider a mental health condition. In other words, some mental health issues may originate from basic immune system processes which initially evolved to help us.

As for how we might act on this, or how to think about this for yourself if you or a loved one is suffering from a mental health condition, it's very important to note that cytokine levels cannot really be used to categorically assess, diagnose or predict any individual person's mental state. As we've already discussed, one reason is that cytokine levels rise and fall depending on whether we're currently fighting an infection, and any given cytokine may affect someone differently, depending on the levels of other cytokines in their body. This means that plenty of people will have higher than average levels of a particular cytokine in their blood and not have any difficulties with mental health. Likewise, plenty of people will struggle with depression or anxiety and not show elevated levels of cytokines.

Where this could be practically useful, however, is if people with a mental health issue have a blood test which reveals elevated cytokine levels, which could be used to indicate the types of treatment that may help them, such as a treatment to block that cytokine. We're not there yet, because a proper clinical trial is needed to test this. Even so, already it's been shown that people with depression and high levels of blood cytokines tend to be those who respond less well to standard antidepressant medication.[41] Still, cytokine levels are not the only thing to affect immune and mental health. Another is the microbiome.

What can the gut-brain-immune connection teach us about mental health?

One especially influential experiment exploring the relationship between the microbiome, immune system and mental well-being used mice that had been genetically altered to lack a part of their immune system (including having no T cells or B cells, and being unable to produce antibodies at all). Their level of 'anxiety' was assessed using a so-called light/dark box test, in which the mice were placed in a large plastic cage with dark and light areas that they could freely access. Mice have an inherent preference for being in the dark, but they also like to explore, which is what sends them into the light areas. In this experiment, anxiety (or something like it) was taken to be reflected by how much time they remained in the dark, choosing not to explore. The results showed that mice lacking a proper immune system spent more time in the dark and explored their environment less. This is an amazing finding, as it suggests that the immune system directly influences mouse behaviours which could relate to anxiety. But that's not all. What's even more surprising is that when probiotics were provided in their drinking water, the genetically manipulated mice behaved like normal mice: they explored their environment more and became less anxious.[42]

We don't really understand how deletion of a major part of the mouse immune system changed their behaviour, let alone how this gets rescued by probiotics,[43] and in any case it's tricky to interpret this in direct relation to human feelings. But it's certainly intriguing and provocative, and adds weight to the general idea that the immune system impacts behaviours and emotions that could relate to mental health conditions – and that the gut microbiome might be involved.

In people, too, the gut microbiome has been linked to anxiety, depression and other mental health issues.[44] People with depression, for example, tend to show less diversity in their gut microbial composition and increased levels of some types of bacteria.[45] Certain microbes in the gut are also associated with attention deficit/hyperactivity disorder (ADHD),[46] and bipolar disorder.[47] Remember, though, that cause and effect are extremely hard to establish, and people with mental health issues may just tend to vary in their microbiomes for other reasons – perhaps difficulty sleeping, levels of exercise or diet. Still, it seems *feasible* that the microbiome could affect our mental health conditions because of its effect on the immune system.

There are other possibilities too. Astonishingly, some gut bacteria produce neurotransmitters – chemical messages for neurons. These could feasibly act directly on the vagus nerve which sends messages to and from our gut and brain.[48] Gut bacteria can even produce the neurotransmitters dopamine and serotonin, both of which are associated with happiness, pleasure and mood.[49] It is feasible, then, that gut bacteria could very directly 'talk' to the human brain to influence how we feel.

Could this be used to help us medically? One very small trial showed that people with irritable bowel syndrome and mild-to-moderate depression who took a probiotic for six weeks diminished their symptoms of depression. MRI scans of those taking probiotics also showed changes in brain activation while looking at fearful or happy faces. These changes happened in areas of the brain considered to be involved in mood regulation.[50] The trial was too small to infer medical advice from, and we don't know enough yet about how the gut microbiome can influence the brain to consider designing probiotics for mental health – so-called psycho-biotics – but this could happen one day.

What about genes — how much of our mental health is determined by immune genes?

Schizophrenia, a long-term and complex condition in which patients may experience hallucinations, delusions and disordered interpretations of reality, is more strongly heritable than depression and many other mental health conditions, so one way of trying to understand its origins has been to compare people's genetic inheritance. Given its symptoms, what's especially surprising is that the genes which have the biggest effect on the likelihood of someone developing schizophrenia are ones that normally work in our immune system.[51]

In detail, susceptibility to schizophrenia is associated with particular versions of immune system genes which enable the immune system to detect the presence of things that are alien to the human body, and play an important role in how an immune response is targeted to an exact type of germ.[52] However, it's not obvious how these genes could affect the likelihood of developing a neurological condition like schizophrenia.

One possibility is that an infection could be involved, at least in some cases, either triggering the condition or amplifying symptoms. In that case, immune system genes show up as important because they affect how well we deal with that infection, which in effect would mean that the link between the immune system and schizophrenia arises from the normal function of our immune system in protecting us against infections — it's just that here a particular type of infection is also a risk factor for a psychosis developing.

Indeed, there is some evidence that infection with a

single-celled parasite called *Toxoplasma gondii* is a contributing factor for the development of some cases of schizophrenia.[53] People can be infected from eating raw or undercooked meat or contact with faeces from an infected cat. The parasite is not usually dangerous, and very often there are no symptoms at all. In the US, about 11 per cent of adults are infected, but intriguingly, people with schizophrenia are more likely than the general population to carry the parasite. We don't know why.

Cats are especially important spreaders of this parasite, although it can infect lots of other animals too, including mice. Of course, mice are normally afraid of cats, and so don't go near them, but astonishingly there is evidence that the parasite has evolved a sneaky process of affecting the mouse brain to make mice less afraid of cats, so they are more likely to pass the parasite on to one.[54] It is possible (but unproven) that the parasite affects the wiring of the human brain too, which could in some vague way – this is very speculative – relate to why some people infected with this parasite also develop schizophrenia.

More directly, the parasite may make some people susceptible to developing a psychosis because of the immune response it triggers. Indeed, there are several lines of evidence to show that an immune response in the body – perhaps specifically one acting in the brain – may increase the risk of schizophrenia developing. First, cytokine levels in blood tend to be higher than average for people with schizophrenia,[55] including two specific cytokines we've come across already, TNF and IL-6.[56] Secondly, post-mortem studies of brain tissue has found higher numbers of immune cells within the brain of people with schizophrenia compared to those without a mental health condition.[57] PET scans show that these immune cells are especially active, in both people with schizophrenia and those with sub-clinical symptoms considered to be ultra-high risk for developing schizophrenia.[58] All of which fits with the immune system directly affecting our mental state, which may be a factor in the onset of other mental health issues too.

Depression is heritable, too, though less so than schizophrenia. If someone has a parent, brother or sister with major depression, their own risk of developing depression is increased by two to three times – it might be 20 to 30 per cent instead of an average of, say, 10 per cent.[59] An analysis of over 800,000 people identified 269 specific genes associated with depression.[60] Thirty-four of these are known to be important in the immune system.[61] One, for example, called olfactomedin 4, has also been linked to gut inflammation. How this may relate to depression isn't known, but we can see that a variety of immune system genes relate to the risk of mental health issues, including schizophrenia and depression.

Are there other ways our immune system affects the human brain?

There is another completely different, and potentially just as important, way in which the immune system can influence mental health. Which is that the immune system can affect *development* of the human brain itself. How is not very well understood, but from what we do know, once again cytokines are vital.

As well as cytokines being produced by immune cells at the site of an infection in the lung, liver or elsewhere in the body, they are produced directly inside the brain. Some neurons produce cytokines themselves, but a greater quantity of them can be produced by special cells within the brain called microglia, which are a bit like immune cells found elsewhere in the body but customised for the brain. Cytokines produced elsewhere in the body during an infection can act on neurons directly, as we've seen – to make us feel sleepy, for example – but they also stimulate the brain-resident immune cells to produce more cytokines there. To what purpose isn't entirely clear, but we do know that cytokines can regulate synaptic communication in the brain to influence our social behaviour.[62] This makes sense in terms of reducing the spread of an infection person-to-person. It's harder, though, to figure out why cytokine levels in the brain can also affect learning and memory. It might just be that cytokines are important in the immune system, and these exact same molecules are repurposed in the brain, without there being any rationale connection – a bit like how you might use the same kind of battery to power this or that device, without them being connected in any way. Nevertheless, the health consequences of the same molecules being used in the immune system and the brain are certainly important.[63]

One study following a group of pregnant women measured their cytokine levels three times, at an early, middle and late stage. Their blood level of the cytokine IL-6 was taken to reflect something about the level of inflammation their body was experiencing at the time. About four weeks after birth, MRI scans were carried out to compare the babies' brain structures. Later, when the children were two years old, they came into the lab to play games that tested their cognitive skills and memory. For example, they had to place stickers inside distinctive containers on a spinning tray; then the tray was spun around, changing the positions of the containers, and they were asked to recall which ones had stickers inside. The results were striking. Blood cytokine levels while pregnant correlated both with the children's brain organisation soon after birth and their ability to do well in a memory game at two. Higher cytokine levels, it was discovered, correlated with poorer memory skills.[64]

This is an extremely provocative result, appearing to say that blood cytokine levels in a pregnant woman can affect their developing baby's brain. But we don't know if or how one thing leads to the other – genes, social interactions and other factors could very well underlie these associations. But on the face of it, this implies an ability of the immune system to act during pregnancy, at least potentially, to affect our lives profoundly.

Another study, this time of over 2 million people in Sweden, found that if a pregnant woman is hospitalised from an infection, the chance of her child having autism is increased by 37 per cent.[65] This sounds worrying, but again caution is needed. This is correlative, so whether one thing leads to another is unknown. More importantly, the actual likelihood of having an autistic child is still very small: the percentage increase sounds dramatic, but the actual level of risk increases from close to 1 per cent to 1.3 per cent. Thankfully, a short-lived immune reaction triggered by a vaccine during pregnancy does not impact childhood development, and this has been explicitly tested for Covid-19 vaccination.[66]

This whole realm of research is complex and difficult. There are lots of intriguing associations, with genes, the microbiome and the state of the immune system, but it's hard to precisely identify the causal arrow. Weighed together, however, most experiments and analyses suggest that the actions of our immune system can affect our mental health to some extent, which begs the ultimate question: what does it mean for treatments, remedies and lifestyle choices?

Can treating the immune system help treat mental health conditions?

Since the mental health of those with an inflammatory disease often improves when they take anti-inflammatory medicine, is this a direct side-effect of said medicine or is their lift in mood a consequence of improvements in the inflammatory disease itself? It's hard to know for sure, without more tests to see whether these types of medicine can help alleviate depression in people with no obvious inflammatory disease. Still, if you've heard that taking a common anti-inflammatory drug like aspirin or ibuprofen can be a quick fix for depression, think again. It doesn't work. One study of young people with depression aged from fifteen to twenty-five showed that aspirin had no effect on their mental health.[67] Another tested whether a low dose of aspirin could alter the likelihood of depression developing in older adults. Again, no effect.[68]

Perhaps the most obvious question is whether blocking a cytokine, exactly as done for patients with rheumatoid arthritis – where we began – helps people with mental health issues, even when they don't have rheumatoid arthritis. The rationale would be that stopping an underlying inflammation, by stopping a specific and direct line of communication between immune cells, might help with mental health. However, where this has been tested, the results so far have either been negative or unclear.

One small clinical trial involving sixty patients found that blocking the cytokine TNF with anti-TNF medicine had no effect on depression,[69] though an improvement for a few people could be teased out of the data by specifically analysing patients with particularly high levels of a blood marker for inflammation. Another small trial showed that anti-TNF medicine did not help patients with bipolar depression,[70] though again a small sub-group of patients did show some improvement. Another drug developed to inhibit immune responses including decreasing cytokine levels also had no effect as an anti-depressant.[71] Overall this type of medicine that blocks the actions of a cytokine doesn't look likely to help mental health problems in a broadly applicable way, but it remains possible that a specific subset of people could benefit. This is perhaps not surprising, because depression is not one singular thing, and neither is any mental health issue. Future research must devise tests to identify people which might benefit: could it be that people with specific mental health conditions *and* higher-than-average levels of various cytokines in their blood (measured a few times so as not to be indicative of a transient infection) and perhaps some other tell-tale mark yet to be discovered, might feasibly be helped with anti-cytokine medicines?

Other big ideas are on the table too. We have already seen that some cytokines stop, amplify or change the actions of others, so it's possible that targeting just one cytokine might not be enough: perhaps we need to use a combination of medicines to block the activity of not one but several different cytokines if we're to influence something as complicated as depression. Also, I and others are wondering about targeting the receptor protein molecules which sense cytokines rather than the cytokines themselves. Could different cytokines, such as those which dial down an immune response, as naturally must happen when infection has been cleared, be used as a medical treatment to help alleviate depression?[72]

The overarching conclusion is that several lines of evidence point to an overlap of immune and mental health. Without doubt, mind and body are not separate entities, in fact they could not be more deeply interwoven. But does this suggest clear strategies for treating or helping mental health? Not conclusively, not yet. In the long term it may well be possible to identify a subset of people under the broad label of any mental health condition who could benefit from medicines that work directly in their immune system. If there is anything practical to take away now, it's that everything we've discussed about how to take care of immune health – reducing long-term stress, getting a good night's sleep, exercising and so on – could help some people with their mental health too. We already know that managing long-term stress, for example, can directly help mental health, but what we see now is that this may work indirectly too, by helping our immune health.

8

100-Year-Long Immunity – or, how does immune health change as we age?

No matter how much we rage against the dying of the light, the body ages, and palpably so on the surface. The upper layer of human skin you can see and touch is regenerated every month or so, but as we age our skin becomes less good at renewing itself. Cuts heal more slowly. Skin becomes wrinkly, because elderly skin cells produce less collagen and elastin, and sebaceous glands produce less oil. Surface bruises happen more easily, because blood vessels in the skin become increasingly fragile.

Beneath the surface the body ages too. Crucially, our defence against infections tends to grow weaker. Yet it cannot just be that the human immune system becomes less reactive as we age, since we are also more likely to succumb to auto-immune diseases like rheumatoid arthritis which result from active, unwanted immune responses. In fact, what happens to our immune health as we get older is anything but simple: many different things are happening all at once.

In evolutionary terms it makes some sense that the immune system works best when we're young.[1] After all, by the time we reach old age most of us have already had any children we were going to have, and they are now independent. In other words, the force of evolution for adapting or improving the human body declines as we age. Add to this that throughout history humankind had an average lifespan much shorter than now, some of which can be accounted for by lower rates of death in childhood, but we are also living longer. In 1950, life expectancy in East Asia was forty-five; today it is over seventy-four.[2] In the UK and US the number of people over ninety has trebled over the last thirty years.[3] So another reason why there may not have been much evolutionary pressure to keep the immune system working well in old age is that for much of human history we tended to die young anyway.[4] If the immune system has evolved to be robust mainly while we're young, what actually happens as we age, how does this affect us, and can we do anything about it?

What happens to immune health as we age?

One big idea of what happens to us as we age is 'inflammaging', a word coined in the year 2000 to denote a combination of inflammation and ageing.[5] Over time, goes the thinking, a low level of background inflammation develops in the body. To unpack this, let's remind ourselves what inflammation is and how it shows up in our everyday lives. The body's response to a cut or a splinter is familiar: once the splinter lodges itself in our body, redness, swelling and tenderness develop. This is inflammation. Under the skin, the damage itself, or the abrupt accompanying appearance of germs, trigger the production of cytokines and other molecules, which in turn attracts swarms of different immune cells to fight off any opportunistic germs that may enter the wound and help with the healing process. Hence the swelling, warmth and redness, as blood flow increases to bring the cells into the contaminated area. This is a rapid, or acute, type of inflammation, which is likely to calm down in a few hours if the wound is cleared of germs quickly and the damage isn't ongoing.

Another type of inflammation is less conspicuous and more enduring. In the joints of people with rheumatoid arthritis, for example, inflammation builds up slowly and hidden from view, eventually leading to symptoms that can last for many years. The same word – inflammation – applies because this process also involves cytokines and other molecules causing an infiltration and activation of immune cells. So while a cut causes inflammation for a short time, inflammation can persist for a long time in an inflamed joint. Though varying in duration, both examples are largely confined to a specific part of the body. However, whole-body inflammation can also arise – the type we discussed in Chapter 3 in the context of obesity, which increases our risk of developing cardiovascular problems or rheumatoid arthritis. This kind of inflammation can be very hard to notice if it is at a low-level, but one tell-tale sign is often raised blood cytokine levels.

The whole notion of inflammaging originates from one simple fact: older adults tend to have higher levels of certain cytokines in their blood than young people, even when no obvious disease or infection is present. As we've seen, the precise levels of most cytokines in blood tend to flux over a twenty-four-hour cycle: for example, the cytokine IL-6 is elevated at night, and associated with us feeling tired. When the body is experiencing an infection, blood levels of IL-6 rise beyond the normal, and in an extreme case like sepsis blood levels of IL-6 can rise by over 100 times.[6] However, the baseline, background level of IL-6 in blood tends to increase anyway as we age, nowhere near as dramatically as with sepsis, but by something like two or three times all the same.[7] Other cytokines also increase in the blood as we age, including TNF, which we have come across as playing a direct role in the joint inflammation which occurs in rheumatoid arthritis.[8] In fact, baseline cytokine levels in blood also rise in elderly animals, reinforcing the idea that there's something fundamental about this.[9] So, what does it mean for our health?

From what we know about the actions of these cytokines, elevated levels will put various types of immune cell on high alert, and though the increase in cytokine levels in old age is not necessarily sufficient to trigger an immune response as such, it is likely to lower the threshold at which an immune response does get triggered. Another way of thinking about this is that as we age the immune system is continuously *slightly aroused*,[10] meaning an immune response might be more likely to fire up when it shouldn't, which could therefore underlie some of the immune health issues we know to occur. If an immune response is more likely to trigger when it shouldn't, this fits with an increased likelihood of an autoimmune disease such as rheumatoid arthritis developing in old age.

Additionally, because as we age blood cytokine levels are higher anyway, when the immune system is needed to fight off an infection the response may be harder to execute. Normally a dramatic and acute rise in cytokine levels from a baseline low level helps mount an effective immune response, but if cytokine levels are already elevated it may be hard for levels to rise further, or sharply, to get the right immune response going. Persistent elevated cytokine levels may also mean immune cells become desensitised, like the boy in the fable who cried wolf. In this analogy the body has been repeatedly signalling that we are being attacked by a wolf (cytokine levels rising), so when an actual wolf does attack it's simply taken as another false alarm and no need to react. This could explain why, as we age, we respond less well to an infection.

There's a further problem. Whenever a real infection clears, the body decreases production of inflammatory cytokines and starts producing a different set of cytokines which act as anti-inflammatory agents. This process is important for helping the body come back down to its normal resting state. You don't want your immune system to stay in the heightened state of activity it needed to be in to fight an infection once that infection has cleared. Otherwise, as we've come across earlier, a fever might persist longer than necessary, you might stay excessively tired, and the immune system might even mistakenly damage healthy tissues and organs. But this whole process of the immune system coming back down to its normal resting state is likely to be more difficult in the context of a heightened background level of inflammatory cytokines that persists on account of ageing. This could also be why, as we age, we take longer to get over an infection, as well as being at increased risk of autoimmune disease.

All this is somewhat speculative – within reason, but not proven – but does fit with what we know about the actions of cytokines and about older people being worse at detecting, responding to and then recovering from illnesses. In effect, the very same immune system processes that are vital for our defences throughout our childhood and most of adulthood have a tendency to become something of a problem later in life. With serious consequences.

In the US, those over sixty-five years of age make up 12 per cent of the population but account for 34 per cent of prescription drugs and 50 per cent of hospital stays.[11] Around 85 per cent of people who die from the flu virus, are sixty-five and over.[12] You'll also remember, I'm sure, that very early on during the Covid-19 pandemic it quickly became clear that the severity of symptoms and risk of dying were worse for older people.[13]

There's another effect that a rise in background inflammation might have on us as we age – not proven, I hasten to add, but possible. In the last chapter we saw how a rise in cytokine levels correlates with an increased risk in mental health conditions, so it is possible that the rise in cytokine levels as we age – a background level of inflammation – may relate to an increased risk of depression or depressive feelings.[14] So much of who we are, or who we become, rippling out from an uplift in cytokines.

Everyone wants to be happy and healthy for as long as possible, and have a handle on how to maximise this, so it couldn't be more important to understand where this uplift in cytokines comes from. Why do cytokine levels increase as we age? One answer comes from another big idea to do with immune health and ageing: senescence.

What is senescence and how does it affect our immune health as we age?

In October 2019, the House of Lords assembled a panel of experts to discuss ageing. On it was Arne Akbar, an immunologist at University College London who has been studying ageing and the immune system for over twenty-five years. 'The focus right now is on senescent cells,' he told the gathering. 'They are the cell of the moment. Everyone who works on ageing is interested in them. The way they are thought to be deleterious is by the inflammatory molecules that they secrete.'[15]

He was right; senescent cells are very important to what happens to immune health as we age. Over a lifetime, cells accumulate in the body that are still alive but have lost their ability to multiply; zombie cells, if you like, or more formally, senescent cells. They accumulate especially in the skin, liver, lungs and spleen, and have both beneficial and detrimental effects.[16] Beneficial because they contribute to repairing damaged tissue, but detrimental since over a long period of time, as senescent cells increase in number, they can disrupt the normal structure of organs and tissues and, as Akbar said, also produce a great many molecules that cause inflammation – namely, cytokines.

A senescent cell is not a particular kind of cell in your body, like a neuron or red blood cell, but something many types of cell can turn into. One way this happens – and probably the most common way a cell becomes senescent – is when a cell becomes damaged. For example, excessive sunlight, or UV light to be precise, can damage a cell's genes. A cell can sense its own damage, and if this happens it may very well switch on its program to become senescent.[17] Why would a cell do this to itself? Because it effectively stops a cell with damaged genes multiplying, thereby preventing it becoming cancerous.

Another way in which a cell can become senescent relates to genetic structures called telomeres. Telomeres act like the plastic tip of a shoelace, preventing the twisted coils of our genes from fraying at the ends or knotting together. But telomeres shorten each time a cell divides, and when they get too short a cell stops dividing and either dies or becomes senescent.[18] Once again, it's thought this happens to protect the body from cancer. This is because every time a single cell divides into two, the newly produced daughter cells gain some genetic mutations from the parent cell, because the process of copying genes into new cells is not 100 per cent accurate, and so limiting the number of times a cell divides decreases the risk that a cell will gain a set of genetic mutations that cause it to become cancerous.[19]

A complication is that some types of cell in the human body, and especially in the immune system, actually *need* to divide lots of times to do their job. For example, the right types of immune cell to fight a particular infection must multiply rapidly many times to be effective. To do so, immune cells can use an enzyme called telomerase to increase the length of their telomeres. But everything comes at a price, and this very same enzyme can also be exploited by cancer cells, which use this enzyme to stop their own telomeres shortening, so they are able to keep dividing and spread.[20] In this way, cancer can develop.

As well as being a simple mark of how many times a cell has divided, the length of telomeres in cells varies from person to person, correlating with certain behaviours and traits. For example, higher physical activity and lower alcohol use tends to correlate with people having longer telomeres in their blood immune cells.[21] In other words, these cells look less aged. Smokers, on the other hand, tend to have shortened telomeres in their blood immune cells.[22] Women who have smoked one pack of cigarettes a day for forty years have telomeres shortened by an amount which corresponds to an extra 7.4 years of ageing,[23] which shows how much of an effect smoking has on pushing cells towards becoming senescent and causing problems, relative to how much this happens anyway as we age.

More generally, the length of telomeres in a person's cells is sometimes used as a measure of a person's so-called 'biological age'. That is, how 'old' your body seems to be rather than just how long you have lived. Even within one person, different organs and tissues seem to have different 'ages', according to the average telomere length of cells, and other measurements too (such as the number of genetic modifications cells have).[24] At the moment it's not obvious how to use this information to target healthcare, but it's possible that knowing the 'ages' of different body parts could be used in the future as a very personalised way of assessing how your various organs and tissues are doing.[25]

For now, what we know is that short telomeres, and genetic damage, lead to cells becoming senescent. Senescent cells tend to produce cytokines and other factors which generate an inflammatory environment in the tissues where they reside.[26] In other words, senescent cells themselves cause inflammation in the body which increases as we age – inflammaging – which in turn affects our immune health.

Given how problematic senescent cells are, why doesn't the body remove more of them? The immune system itself is indeed responsible for clearing out the body's senescent cells,[27] but eventually it can't keep up with the task in hand. The problem is amplified by senescent cells being able to induce senescence in other cells nearby. It's not known for certain why this happens, but one idea is that, as well as helping protect us from cancer, senescence might help control viruses, since a virus will find it harder to replicate itself inside a senescent cell.[28] Indeed, cells can become senescent when they are infected with some types of virus (such as respiratory syncytial virus or RSV and Epstein–Barr virus), presumably as part of our defence against them.[29] In that case, it might also make sense for an infected cell which turns senescent to cause a zombie-like state in nearby cells so the virus can't easily spread and multiply.[30] But if senescent cells generally make adjacent cells senescent, this adds to the problems they cause. As convoluted as this sounds, senescence is at once a problem for immune health and a part of our immune defence.

Another complication is that immune cells can themselves become senescent. As we age, many blood-borne immune cells show signs of senescence, and immune cells in our blood that have become senescent are particularly adept at producing cytokines even when there's no *bone fide* threat for them to be active against.[31] This, we know, tends to happen more from around the age of fifty.[32] It's not known why this age in particular may represent something of a turning point, but we do know that one particular set of immune cells are especially prone to becoming senescent: memory immune cells.

Each time we fight an infection our body retains some of the immune cells best equipped to tackle that infection – memory immune cells – in case we encounter the same germ again. These long-lived cells account for the body's ability to fight off infections more quickly a second time round. They are also important in enabling vaccines to work. But as we age, the number of memory immune cells in the body increases from us having already spent decades battling germs. You would think this helps us, and absolutely it does, to stay safe against germs we've experienced before, but surprisingly two problems develop as well. First, as the bank of our memory immune cells increases, there is a corresponding reduction in the number of immune cells we have that are able to respond to new infections.[33] It's not completely clear why this happens, but feasibly the total number of immune cells in the human body is limited. This could be why an older adult is not as good at reacting to new infections. Secondly, as with all cells, some memory immune cells inevitably become senescent, and these will pump out cytokines, albeit at a low level, creating a background inflammation, which simply isn't good for us.[34]

A further problem with senescent cells affecting our immune health occurs in bone marrow, the soft, spongy tissue running through the centre of bones, which is both surprisingly complex and surprisingly important for immune health. It contains blood vessels, nerves and a wide variety of stem cells, which generate all sorts of other cells for the body, especially immune cells. As we age, however, a significant fraction of stem cells in bone marrow become senescent (or something very similar). This is one reason why cancer charities which recruit bone marrow donors for transplantation are especially keen to sign up young people.[35] Normally, stem cells in bone marrow continually produce a steady stream of new immune cells to circulate in our blood, especially B cells (which in turn, produce antibodies), but this works less well as we age.

As well as a direct effect on stem cells, another significant change happens in bone marrow: fat-storing cells increase in number. These cells produce a number of different types of molecule (protein and fat molecules, for example) which tend to skew the production of new immune cells towards one specific type.[36] We don't know why, but we do know that these alterations in bone marrow – the decreased regenerative power of stem cells, and increase of fat cells – almost certainly relate to why older adults tend to have fewer of some types of immune cell in their blood,[37] and underlie the change in relative levels of other kinds of immune cells.[38] The exact consequence for our immune health is unclear, but seems likely to be detrimental.

On the whole, then, senescent cells cause problems as we age.[39] To test whether their removal can directly benefit health, we must turn to experiments with animals. To this end, mice have been engineered so that whenever a cell in their body turns senescent, a gene is activated to make the senescent cell capable of being killed by a drug treatment and we can assess the consequences.[40] The result was dramatic: a lifelong removal of senescent cells delayed many signs of ageing. The mice developed fewer cataracts, muscle fibres were thicker and fitness levels were hugely improved.[41] These results came from eliminating senescent cells throughout life, but another experiment tested what happens if mice were left untreated until midlife. Compared to mice whose senescent cells were left alone, these mice had kidneys that worked better, their hearts were stronger and they tended to develop cancer at a later age. There were no apparent side-effects. They even showed signs of having better mental health, in that they tended to explore their cages more. Perhaps most dramatic of all, their lifespan extended: on average, they lived 17 to 35 per cent longer.[42]

Of course, this elimination of senescent cells in mice can't be replicated medically in humans, because it involved a specific genetic alteration in the mice to make senescent cells susceptible to being killed by a drug. Nonetheless, this research is important because it means that drugs which can selectively destroy senescent cells are probably something to aim for. But that's not to say it's easy. Senescent cells are human cells without anything obvious to distinguish them, at least as far as we know, and so are hard to target, especially in a way that would leave healthy human cells untouched.[43]

Towards this goal, however, a handful of genes have been identified that can be individually deleted to have the precise desired result of making only senescent cells die, not healthy cells. Many of the genes that can do this turn out to play a part in senescent cells introducing large amounts of protein molecules, including cytokines, into the liquid surrounding them, to affect other cells.[44] Evidently this process is important not only in senescent cells exerting their effect on the body, but also to senescent cells staying alive in the first place. A weakness, and a target. We don't yet have drugs to exploit this vulnerability, but one day we might, and then maybe we'd have a medicine that can safely eliminate senescent cells, prevent some aspects of ageing and restore immune health in older adults.[45]

What other things affect immune health in old age?

Ageing is multi-dimensional, and it would be wrong to think its effects on immune health are only related to inflammaging or an accumulation of senescent cells. Other things happen too. For one, there tends to be a shift in the microbiome as we age, with the bacteria which produce butyrate and other molecules decreasing in abundance. As we saw in Chapter 2, butyrate is important in regulating the immune system[46] by directly affecting the activity of T regs, the immune cells whose job is to stop unwanted immune responses. This could very well be another reason why, as we age, our immune system becomes less regulated, and more inclined to inadvertently attack normal healthy cells and tissues, which can give rise to an autoimmune disease.

Another influence on immune health as we age is the thymus, a small organ which sits in the chest between the lungs and is important to the body's production of T cells.

Over time, the thymus shrinks, but with what consequences for our immune health?

To answer this, we first need to look at what happens when new T cells are produced. A shuffling of genes gives each individual T cell a uniquely shaped T cell receptor molecule which protrudes from its surface and determines what it will react against. Crucially, the reactivity of each T cell develops randomly, so all kinds of T cells are produced with all kinds of reactivity. In total, the human body is thought to have around 4 trillion individual T cells,[47] only a small fraction of which will have the right-shaped T cell receptor able to detect the presence of any particular germ. And because T cells are produced with random-shaped T cell receptors they have an ability to lock onto germs the body has never seen before – even germs that have never existed in the universe before. But this means that some newly produced T cells could react against your own body's healthy cells or tissues, so before any T cell is let out to patrol the body, it is examined in the thymus to check it won't. Any which do are killed off. This effectively guards against new T cells patrolling the body and reacting against the body's normal healthy cells and tissues, which would cause an autoimmune disease.[48]

For me this process is one of the most wondrous things in the entire human body: the way in which the immune system can manage to protect us from *any* kind of germ while also stopping itself attacking our own bodies.[49] This is not something the human mind can intuit. But what's important for our discussion of ageing is that the thymus gets smaller as we age. It's not very clear why, but it happens in all sorts of other animals too (those which have a thymus), suggesting it's an ancient and conserved process.[50] What is clear is that as we age, the body is less able to produce new T cells.[51]

The exact effect on our immune health is not entirely understood, but it could be another reason why older adults are less well equipped to respond to new infections. Which means that to improve immune health as we age, we may want to regenerate the thymus, which takes us into the realm of science fiction, but there are possibilities: hormonal therapies, chemicals which prevent degeneration of the thymus or transplantation of thymus-like tissue that has been cultured in a lab.

All this has to do with changes in the human body itself, but something else changes our immune health as we age: the history of infections we've been exposed to over the years. As we've already seen, after a bout of chickenpox a small amount of the virus remains dormant inside your body, and occasionally it can reawaken later in life and lead to shingles. Shingles is much more likely in older adults; it's thought that when we're young the virus is kept in check in part by an active immune response.

Other viruses have an especially long-lasting impact on our immune health.[52] One is cytomegalovirus, or CMV, a common infection related to the virus which causes chickenpox which, also like chickenpox, infects many people in childhood. Worldwide, about 45 to 100 per cent of adults are infected, with variation country-by-country, and even state-by-state in the US.[53] Most commonly, becoming infected with CMV triggers a strong immune response so that there are no symptoms apart from sometimes a sore throat or fever. On rare occasions, however, something more dangerous can develop in people with a weakened immune system (such as those infected with HIV), and if a pregnant woman is infected there can be serious problems should the virus be passed on to the baby in the womb. Sadly this is a leading cause of neurological and developmental problems in children.[54] But even when the active virus is well controlled by a person's immune system and no clinically difficult situation develops, the virus is never eradicated. It persists in a dormant form, hiding in our bone marrow cells, maintaining a lifelong relationship with the human body. Many people therefore harbour the virus unknowingly, and it affects their immune system.

All sorts of health metrics, including the numbers of different immune cells in blood and cytokine levels, correlate with whether or not someone carries CMV.[55] For example, people over the age of sixty have a higher number of a specific type of NK cell in their blood according to whether or not they carry CMV.[56] What does this mean for our health? It's not that CMV simply weakens our immune system – in fact, it's possible that a mutually beneficial co-existence has evolved between humankind and this virus. Dormant CMV in young adults correlates with them having a stronger immune response to a flu vaccine.[57] On the flip side, there are rare complications, as we've mentioned, and CMV infection has been associated with an increased risk of cardiovascular problems.[58] That CMV affects us so deeply, and for so long, might be a particularly extreme example, but other viruses almost certainly have lifelong effects on us too; we just know a lot less about them.

Our history of infection is something that varies greatly from person to person, which must be a significant reason why young people's immune systems are much more alike than older people's.[59] There's nothing practical we can take from this – it's not even clear what's detrimental and what might be beneficial – but it is another dimension to what happens to our immune health as we age.

Life comes at us fast... How do the big milestones impact our immune health?

Ageing is far more than biological infections and anatomy; it is also an accumulation of experience and wisdom. The older we get, the more likely big life events like the death of a partner or spouse, are to happen to us, or have happened already. A landmark study reported in 1963 followed 4,486 widowers aged over fifty-five. After the death of their loved one there was a 40 per cent increase in the probability that they would also die within six months.[60] There are many possible reasons, some practical, such as married or partnered couples being likely to share lifestyle factors like dietary habits or access to healthcare that would predispose them to similar lifespans.[61] But bereavement also has a direct bearing on immune health.

People who have recently lost a loved one tend to have a weakened response to vaccines, as well as an elevated level of background inflammation.[62] Blood levels of the cytokine IL-6 tend to be higher in bereaved adults than in married or partnered adults of similar age.[63] While there is no correlation between cytokine levels and whether a person is experiencing 'complicated grief' – a longer than normal, heightened state of mourning – what does correlate with strongly elevated levels of IL-6 in bereaved adults is a particular genetic variation.[64] We don't know how, but it does mean some people are naturally more susceptible than others to their health being affected by their loss.

Other life situations impact immune health too. Older adults caring for a spouse with dementia also show a weakened vaccine response[65] and elevated levels of cytokines in their blood.[66] This could all relate to stress, disturbed sleep and less exercise, in which case we can take action to help this. Older adults assigned to a t'ai chi class for three forty-five-minute-long classes a week, for example, showed improved immune responses against the chickenpox virus, tested in a lab dish.[67] Another study found a t'ai chi class led to elderly adults responding better to a flu vaccine.[68] Both studies involved small numbers of people, however, so we should exercise caution, and the findings are hard to interpret in terms of how well someone would fare with an actual infection.[69]

State of mind may be especially important to immune health in old age. Older adults who are optimistic and positive tend to spike less cortisol in stressful situations; one analysis of 135 over-sixty-year-olds across six years found that those with an expectation that things would turn out well were less likely to show high levels of cortisol on days when they said they felt stressed.[70] There's less evidence that a positive attitude can help counter the effects of long-term stress,[71] but optimism does correlate with a longer life span and a greater likelihood of living past eighty-five.[72]

I've always considered the kind of positive-thinking homilies and affirmations one comes across on bookmarks, screen savers and the like to be bland and superficial, but it turns out maybe I was wrong. There could be something profoundly important for our health in simply being appreciative of life. One study that has become hugely influential had people with neuromuscular disease reflect every day on things to be grateful for. They went on to sleep better.[73] It's hard to prove, but certainly possible that positive thinking or showing gratitude can buffer stress and help immune health, especially as we age. When I tell my wife I'm writing this she smiles and says, told you so.

So how long can we live for, and what can we do to age well?

Some effects of ageing are caused by things unique to each of us like our personal history of infections, big life events and so on, which makes what happens to us in old age complex and fuzzy; we are not dealing with a physical law of the universe. Which says that at least some of the effects of ageing do not seem inevitable. Could we even reverse or stop the effects of ageing, then, by among other things improving our immune health? Is there a way to avoid the decay and frailty that come with age – to live better for longer?

So far, our quest to understand ageing has led, perhaps unexpectedly, to the realisation that there is no hard or clear biological limit to how long we might live. In many parts of the world life expectancy continues to rise, and nobody knows exactly how long humans might live for. According to the *Guinness Book of World Records*, Jeanne Calment, who died in France in 1997, lived to the age of 122, outliving her grandson.[74] Although he wasn't as lucky as her, long lifespans like this do tend to run in families.

Studies of twins appear to show that the genetic contribution to longevity is around 20 to 30 per cent,[75] and genes seem to be especially important in living beyond the age of ninety.[76] Astonishingly, genetic mutations in nematode worms, which normally live for about three weeks, can increase their lifespan by up to ten times,[77] though needless to say nothing like this is possible for humans. In humans, hundreds of gene variants are linked to ageing, each having a small effect, but combining to have a significant effect overall.

Amazingly, it is now possible to manipulate genes in a lab dish to make cells become young again. In 2005, the Japanese scientist Shinya Yamanaka showed that the introduction of four genes – the Yamanaka factors – into adult cells caused them to revert to stem cells normally found in embryos, capable of becoming all the different kinds of cells in the human body.[78] For this, he was awarded a Nobel Prize. Today researchers are seeking ways to control this process more finely; to use Yamanaka factors to roll back the age of cells, or repair damaged tissues, without going all the way back to an embryo-like state. We have, for example, altered a fifty-three-year-old woman's skin cells so they behaved as though they were thirty years younger, by exposing aged skin cells to Yamanaka factors for a relatively short period.[79] Significant problems remain, because the very same factors that make cells young again can also contribute to cancer. As we search for eternal youth, therefore, problems arise from the fact that cancer is what happens when cells become immortal.

Nobody knows the extent to which humankind is going to conquer ageing. But it is clear that studying immune health and ageing offers the opportunity to understand a lot of hugely important issues, with exciting discoveries and spin-offs certain to emerge, such as novel vaccines or new ways of fighting cancer or assisting tissue repair. The mission to improve our immune health as we age has a vastness to it in the same way that landing on the Moon wasn't just about landing on the Moon: it is a journey that will lead to all manner of new technology, scientific knowledge and practical outcomes.

The nematode worms provide a note of caution. The

long-lived genetic mutants also ended up with a much-extended period of frailty, underlying the importance of increasing health-span, not just lifespan. And we must also consider the health of society, because ageing is not just biological, but intertwined with social, economic and psychological circumstances that mean improved health in older adults has far-reaching consequences. Should we work until our seventies or eighties? How will we ensure equality when even now the rich live longer than the poor? Is euthanasia something we will have to consider more and more if we end up living longer and longer, especially if some of our physical or mental abilities simply can't keep up? These are tough questions, that require us to tackle ageing on multiple fronts. The advancement of science has long forced us to address complex social issues, however, and arguably we are making better progress in the biological science itself than in the issues which come from it.

In the long term it is very likely we will find new medicines that help, perhaps by eliminating senescent cells or lowering background inflammation. In the meantime, as we age, we're on a tightrope. Small nudges to keep us balanced are vital, from managing stress to keeping active, getting a good night's sleep, even just being positive about life (which seems a low bar but is sometimes the highest). We are ensorcelled by the beauty of youth, but we must work harder to make old age palatable, fulfilling and beautiful – and above all healthy. And as we achieve this, each of us must find our own answer to the most pertinent question of all. What will be your purpose, or what would make you happy, in your extra years?

9

In the Pipeline – or, what big new ideas are on the horizon?

One day in February 2018, I was in a South London studio talking to the presenter Eamonn Holmes for his drivetime radio show about how the immune system works. It was pre-pandemic, when the immune system was not often talked about in the media, let alone on drivetime radio, so hats off to the producer for booking me. What things work well for boosting your immune system? the presenter asked me. I replied: vaccination. Eamonn Holmes looked at me as though I hadn't understood the question and the clear implication that he had meant something *natural*. I told him I could see what he was driving at, but one thing that boosts your immune system is vaccination – and to me it's not quite right, or at least too simplistic, to think of vaccination as being entirely unnatural.

What exactly do we mean by a 'natural' way of improving immune health? And what exactly is being considered? Whether something is safe, and does this depend on its source, or merely if it's been produced for financial gain? Of course, we think of a vaccine, an antibody treatment or an antiviral drug as hugely different from, say, a vitamin tablet, a calming candle perfume or anything like that. This makes sense: they are. But equally, there is no sharp dividing line separating what is natural from what is unnatural.

Take penicillin. In 1928, Alexander Fleming noticed that

mould which happened to be growing on a Petri dish had prevented bacteria from growing around it.[1] The substance responsible – penicillin – was isolated and, on 25 May 1940, used to save the lives of four mice deliberately infected with deadly bacteria.[2] In February 1941, a forty-three-year-old policeman, Albert Alexander, became the first person to be treated with penicillin. But we still tend to think of penicillin as a pharmaceutical product rather than something occurring naturally. Vitamin C, on other hand, is seen as natural, because it is found in the food we eat. Both penicillin and vitamin C are chemicals that come from a natural source, both can be produced synthetically and given as a measured dose in tablets, and yet penicillin is thought of as far more 'medical' than vitamin C. But the assumption that vitamin C is more natural than penicillin is nowhere near as reliable as we think.

Oranges are well known to be a source of vitamin C, and most of us would consider them to be natural. But the oranges we eat nowadays are the outcome of centuries of horticultural selection,[3] and to make it available throughout the year the orange juice we drink has usually been separated, filtered, pulped, pasteurised, processed and stored in vats that have the oxygen sucked out.[4] Orange oil from the peel, and the pulp that has been separated from the juice, together with other additives, perhaps, to create a so-called flavour pack, can then be added back in to the orange juice to improve or modify the flavour, all while retaining a 'not-from-concentrate', natural-sounding label. In some tablets or supplements, vitamin C has been derived from a particular type of cherry, acerola.[5] Other tablets or supplements contain vitamin C that has been produced synthetically. The bottom line is that it's very hard to say what's natural. At the level of atoms, vitamin C is the same whether it's come from an orange, orange juice, been isolated from cherries or produced synthetically. But are all these permutations equally natural, and does it matter?[6] How each of us approaches this – and for that matter, anything in this book – in our everyday lives will be not only scientific but also a reflection of social, psychological, educational and economic factors.

Let's take another example. Both whooping cough vaccine and a probiotic yoghurt drink use as their active ingredient bacteria or parts of bacteria.[7] Of course, having an injection of whooping cough vaccine certainly feels more of an intervention than drinking a bacteria-laced yoghurt drink. But some vaccines can be taken orally – polio vaccine, for example – so we can't categorically state that injections are what makes vaccinations seem so unnatural.

Vaccination, then, is a thoroughly emotive subject, and you will have your own views. But it is something of a myth to think there is a clear line between things we can say are natural, and therefore presumed to be safe, or more likely to be safe, and things that are unnatural, which we must necessarily be more wary of. As the science of immune health progresses – as bacteria-laced yoghurt drinks become more complex and more medicinal, for example – taking something to be natural, and intrinsically not harmful, is set to become more difficult.

What are vaccines, and how do they work?

The concept of vaccination has its roots in traditional medicine and folklore. It has been claimed that inoculation – deliberately exposing someone to a small dose of material which is actually infectious – goes back to a Taoist or Buddhist monk around the year 1000.[8] Though this is somewhat speculative, it has been established that inoculations in China and India were carried out in the sixteenth century. Later, in the eighteenth century, it is documented that children in the UK gained protection from smallpox by wearing smallpox-infected clothes, or from smallpox scabs being held for some time in a child's hand. This would lead to mild symptoms developing, and protection against worse symptoms arising from a second infection. Two doctors from Wales, writing in 1722, described this practice as common.[9] Modern vaccinations work on the same principle, but are subtly different to inoculation, because the ingredients of a vaccine are not themselves infectious.

The basic idea is familiar: vaccines work by getting your immune system to react to a dead germ, or an isolated component of a germ, in order that some of the immune cells involved live on in your body for a long time, allowing you to respond quickly if the real germ is ever encountered. But, truth be told, vaccinations are actually more complicated than this, and a little bit mysterious too.

The process by which they work as just described is too simplistic – that by exposing your body to molecules from a germ primes the immune system to react rapidly in due course to the germ itself. If that were all there was to it, then it ought to be straightforward to create vaccines against any kind of germ: we'd just inject protein molecules from the germ into people and trigger an immune response to give them protection from the actual germ. The immune system, however, is so complex that when a new vaccine is being developed it is extremely hard to predict whether it will work, and even if it does, how long it will protect us for. How come we could create a vaccine for Covid-19 so rapidly, but still not have one that's effective against HIV?

It is amazing – and game-changing – that several vaccines were developed against the virus that causes Covid-19 within a year. In Oxford, in January 2020, Sarah Gilbert's team designed what would eventually become known as the Oxford-AstraZeneca vaccine over just a few days. Sixty-five days later and the first batches of the vaccine were to hand, and they had the results to prove it worked from a huge and well controlled trial on 23 November that same year. Others created vaccines too, also rapidly. The 2023 Nobel Prize in Physiology was awarded to Katalin Karikó and Drew Weissman for their role in creating messenger RNA, or mRNA, vaccines. Altogether it has been estimated that in the first year of vaccines against Covid-19 being used, almost 20 million lives were saved.[10] By contrast, it has been more than thirty-five years since AIDS was first identified as being caused by a virus, and we still do not have a vaccine.

There are numerous reasons why vaccines arrived so rapidly for Covid-19. The genetic make-up of the virus was determined quickly, and thanks to a good understanding of related diseases we weren't starting from zero.[11] More generally, decades of research developing vaccine technologies meant that a large number of research teams, each using a different approach, could hit the ground running.[12] Volunteers readily stepped up for tests, and because the virus was so prevalent it didn't take long to see whether the vaccine offered protection. Finding a vaccine became a global priority, so government bodies whose job was to assess data to decide which new medicines should be licensed for general use put Covid vaccines at the head of the queue. Above all, perhaps, it was fortuitous that the virus causing Covid-19 happened to be the sort of virus for which a vaccine can work well. It's one seen relatively easily by the immune system, and vaccinated people usually produce plenty of antibodies that can recognise the virus and prevent it from infecting human cells.

Why, then, don't we have a vaccine for every infectious disease, including HIV? The human immune system is capable of fighting HIV, so a vaccine should be possible, at least in principle. Some people's immune systems, indeed, are especially good at fighting the virus – take the example of woman #1 (aka Loreen Willenberg) in a study of HIV-infected individuals, who has spoken widely about her experience. Loreen was diagnosed as being infected with HIV back in 1992, but has never progressed to full-blown AIDS. Amazingly, her immune system has managed to fight off the virus to the extent that by 2020 it was virtually undetectable in her blood.[13] This woman is very special, but her situation is not unique; around one in 300 people infected with HIV does not progress to full-blown AIDS for seven years or more because, like Loreen, their immune system is able to fight the virus effectively.[14] What this means is that the human immune system can keep HIV under control, and so we need to find a way of getting this to happen in more people.

Which brings us to the fine print of what it takes to make a vaccine work. It's not as simple as injecting someone with a component of a germ, because a protein molecule isolated from a germ isn't seen by the immune system as anything dangerous that warrants a reaction. This crucial revelation has its roots in research carried out in the 1920s by two scientists working independently, the French biologist Gaston Ramon and London physician Alexander Glenny. They found that a protein molecule called diphtheria toxin, produced by the bacteria which cause diphtheria, could be inactivated using heat and a small amount of a chemical called formalin. This inactive form of diphtheria toxin, they postulated, could then be used as a vaccine against diphtheria. But when they injected it into animals it didn't work. The animals were not protected.

However, in 1926 Glenny's team stumbled on the discovery that when inactive diphtheria toxin was combined with aluminium salts, it did work as a vaccine against an actual diphtheria infection. Glenny's explanation was that the aluminium salts helped the diphtheria toxin stay in mice long enough for an immune reaction to develop, but we now know this isn't so. What's actually happening here is that the presence of aluminium salts tricks the animal's body into thinking something dangerous is present and mounting an immune response. Other substances such as paraffin oil, or squalene which is an oil naturally produced by human cells, have since been discovered to help vaccines safely work in the same way that aluminium salts do, and are used in some flu vaccines. A particular type of fat-based molecule originally derived from bacteria can also help initiate an immune reaction, and is used in one of the vaccines against human papilloma virus (HPV). Collectively these chemicals are known as adjuvants.

Adjuvants help vaccines to work because the immune system must be tricked into reacting as though something dangerous has entered the human body when in fact it hasn't. The new mRNA vaccines are a revolutionary type of vaccine, and for these the mRNA itself can act as an adjuvant. To understand how, we need to examine how mRNA vaccines actually work. mRNA is a type of molecule normally found in cells that has the job of copying the DNA sequence of a gene. The DNA itself is kept in a cell's nucleus, but the mRNA copy travels outside that nucleus to the machinery which produces protein molecules. A vaccine, therefore, can use an artificially produced mRNA to instruct the body's own cells to make a protein molecule from a germ. This happens only transiently, because the mRNA is soon broken down – which means the vaccine itself doesn't necessarily linger in the body for long. Importantly, the artificially added mRNA can itself trigger an immune response against the protein it has produced. This can occur because some of the artificial mRNA may spontaneously assemble into a double stranded helical shape which looks to the immune system like a potential threat, even though it isn't, because it is a shape of RNA found in some viruses.[15] In fact, one of the important steps in making effective mRNA vaccines was working out how to avoid eliciting too strong an immune reaction against the mRNA itself, so it would not be broken down too quickly.[16]

In 2005, Karikó and Weissman discovered a way of doing this.[17] A chemical modification to the mRNA did the trick. Amazingly – in hindsight – their paper describing this was rejected within twenty-four hours by the prestigious journal *Nature* with the comment that the discovery was too 'incremental'.[18] Yet these modifications proved in due course to be incredibly important and were used in both the BioNTech and Moderna Covid-19 vaccines to save millions of lives.[19] With the immune reactivity of the mRNA itself dialled down, it turns out that adjuvants are still commonly used in mRNA vaccines: fat molecules packaged with the mRNA act as an adjuvant to kick-start an immune reaction.[20] Generally the choice of adjuvant is decisive in its own right, because they are not all equivalent. Picking the right one helps direct an immune response to fit the type of germ being targeted.[21]

Aside from needing something to trigger the right immune reaction, there are many more hurdles to overcome in making a good vaccine. There have been over 250 vaccine trials for HIV, and none has proved clinically effective, apart from one which showed a hint of success by reducing HIV transmission by about 30 per cent.[22] One reason is that the genetic make-up of HIV varies enormously from person to person, and even within a single individual. Also, in people carrying the virus there tends to be a set of HIV-infected cells that lie dormant, not actively producing new virus, which can then switch on later. This so-called latent reservoir is tricky for the immune system and anti-viral drugs to target.[23]

Another problem is that HIV itself has evolved some very specific ways of avoiding an immune response. One example (which I happened to have played a small role in uncovering) is that HIV directly interferes with how immune cells called T cells detect signs of disease.[24] Normally T cells survey the surface of other cells looking for anything which shouldn't be there, as a tell-tale sign that the cell might be infected. Microscopic fingers protrude out from the ball-like T cell to scan the surface of another cell close by:[25] if a T cell detects something that isn't yours – a piece of protein molecule from a virus, for example – then it will attack and kill the virus-infected cell, as well as send out secretions to alert other immune cells. But HIV interferes with this whole surveillance operation, by stopping the cell it infects from being able to display anything which might alert a T cell. These idiosyncrasies are almost certainly why vaccination to protect us from AIDS has proved so difficult, in contrast to what we've seen with Covid-19.

Compared to the olden-day practices of inoculating children by having them wear clothes from someone infected, vaccines have evidently become much more of a pharmaceutical product. This is for the better, because we now have vaccines to all sorts of diseases, saving many millions of lives every year. (It's also not clear how bad symptoms could get when children were exposed to smallpox under the traditional custom.) But the complexity of modern vaccines – the use of adjuvants, the sheer novelty of some of them, their ubiquity and so on – makes it incumbent on us all to be informed about their nature and how they work,[26] ideally by some source other than solely the pharmaceutical companies manufacturing them.[27] Because vaccinations play, and will continue to play, a very important role in our immune health.

What new vaccines might we expect to see next?

When we go for a vaccine we hope our immune health will be improved for a lifetime rather than just a few months or years. Some vaccines only work for a short time on account of the virus itself changing, as in the case of new flu vaccines being needed annually. But more generally one of the greatest challenges driving new vaccine development is ensuring that any immunity is long-lived. We don't yet understand how to make immunity last many years, let alone a lifetime.

There are also many diseases for which we don't have vaccines at all yet, or at least good ones. Not only HIV but also malaria and cancer have proved challenging. Malaria is well-known as a life-threatening disease caused by a parasite spread by some types of mosquito. One reason malaria is complicated to tackle with a vaccine is that the parasite goes through a series of shape-shifting stages during its life cycle, back and forth through mosquitoes and humans. After decades of research, the first malaria vaccine was finally approved by the World Health Organization in 2021. Eighteen million doses have since been allocated to twelve countries. This vaccine targets a protein molecule found on the parasite when it transmits from mosquitoes into the human bloodstream. Antibodies produced by the vaccine can help prevent the parasites then going on to infect the liver. This vaccine will save lives, but is not perfect because its prophylactic effect is short-lived.[28]

To hit the parasites as soon as they enter the bloodstream, protective antibodies need to be there at a high level. Otherwise, within ten to fifteen minutes the parasites can reach liver cells and change their state so that they no longer have the target molecule on their surface, which is very problematic, as this is all the vaccine-produced antibodies can act on.[29] So as antibody levels wane over time, as we see with a lot of vaccines, protection is lost. Another problem is that children who are protected still carry parasites in their blood that can infect mosquitoes, which can then go and infect others in the population who haven't been vaccinated. While an individual can be protected, therefore, transmission within the community is not reduced.

In 2023, a second malaria vaccine was formally approved, which targets the same parasite protein molecule but is cheaper per dose, meaning it should be easier to distribute even more widely.[30] It's likely that it will also only offer short-term protection, so malaria vaccines still need to be used alongside other measures, including bed nets treated with insecticide, good diagnostics and anti-malarial drugs. Other strategies on the table include genetically altering mosquitoes, using what is known as a gene drive, to make offspring sterile and reduce the numbers of mosquitoes able to carry the malaria parasite.[31]

One new category of vaccine on the horizon uses what is called self-replicating RNA, which is different from the standard mRNA vaccine we've already come across. In this case, RNA from an actual virus is re-used, but in an altered form, so its ability to replicate itself inside cells is left intact, but it cannot become infectious, because the parts which allow it to spread from one cell to another have been deleted. RNA needed to produce the vaccine target, such as the Covid-19 spike protein, is fused with this repurposed viral RNA. It's hoped this will be an improvement on current RNA vaccines because production of a target protein molecule will be high, on account of the RNA multiplying itself, even if a low dose of vaccine is given. This could mean less chance of side-effects, while still triggering the desired immune protection. Indeed, side-effects from current RNA vaccines are not insignificant: over half of recipients experience a headache, and around one in six gets a fever.[32] Many trials of self-replicating RNA vaccines are underway,[33] and this type of vaccine might one day be something we come across frequently as a better way to vaccinate against flu, hepatitis B, Covid-19 or anything else, with fewer side-effects.

New vaccines for cancer are in prospect, too. Since cancer is usually caused by our own cells gone awry, there's nothing as obvious to target as, say, the Covid-19 virus's spike protein. An exception is human papilloma virus (HPV, as it's commonly known), the infection that is tested for in smear tests. In most cases, this virus is cleared by a person's immune system, often without them even being aware of it, but occasionally it can cause cervical cancer; in fact, nearly all cases of cervical cancer are caused by it. This happens if cells harbouring the virus develop abnormalities and spread to tissues below the surface.[34] For this type of cancer specifically, a vaccine has, thankfully, been available since 2006. It works by containing protein molecules from the virus, plus adjuvant to stimulate the immune system to produce antibodies that can stop an infection. This vaccine is currently being used in 120 countries worldwide, and is thought to be saving around 300,000 lives a year.[35]

In general, however, to come up with a vaccine against cancer requires targeting something much more subtle – specific mutations which have arisen in cells to cause them to lose their normal self-regulation and become cancerous. This varies with every individual because there are lots of ways in which cancer could develop, meaning one person's cancer has a particular set of changes from their normal cells that are unlikely to be the same in someone else's cancer. One approach to making cancer vaccines, therefore, would be to analyse a person's own specific cancer, work out which mutations have happened and design a bespoke vaccine to trigger the immune system into targeting those exact changes. This might be done best with mRNA vaccines because it is technically relatively straightforward to change the target of an mRNA vaccine to produce something personalised. In practice – in the future – it might take something like a month or two to produce a personal cancer vaccine after a sample has been collected from a patient.

This was always part of the original mission for mRNA vaccines, in fact. In July 2013 Katalin Karikó gave a talk at the biotech company BioNTech about her mRNA research (before she became famous and the winner of a Nobel

Prize – indeed, she had recently lost her university lab space on account of not getting research grant money). One of the co-founders of BioNTech, Uğur Şahin, then took her to lunch. It wasn't to discuss coronaviruses – Covid-19 was years away; it was vaccines for cancer he wanted to talk to her about. Şahin reasoned that radiation, chemotherapy and surgery were all one-size-fits-all options in treating cancer, and that patients could benefit from something more bespoke. And Karikó and her mRNA research, he believed, could contribute to achieving that. They got on well, bonded by science and hard work: 'like me', Karikó later recalled, Şahin 'didn't seem to distinguish between his work and the rest of his life. His life's work was just that: his life *and* his work.'[36]

It's no exaggeration to suggest that this lunchtime chat shaped the whole destiny of humankind. Because Şahin offered Karikó a job, she took it, and things went from there to us having mRNA vaccines for Covid-19, saving millions of lives across the globe. Now BioNTech and other pharmaceutical companies have cancer in their sights. In July 2023, buoyed by the success of the mRNA vaccines against Covid-19, Moderna and Merck announced the launch of a major trial of a personalised vaccine for melanoma, a most serious kind of skin cancer (and the fifth most common cancer in the UK). Melanoma is a good first test case for new vaccines because it is a type of cancer we know is susceptible to other kinds of therapies that work with the immune system,[37] which means we already know that boosted immune responses can be effective in beating melanoma in some people. But it will take time before we see whether long-term survival rates can be helped by personal vaccines. Final results are due in 2029.[38]

Personalised vaccines are not currently available to cancer patients and, realistically, staying with what we can be sure about with immune health, we don't yet know if vaccination against cancer will work in a broad way (beyond the specific example of the virus that causes cervical cancer). For one thing, the protein molecules that would be targeted by a personalised cancer vaccine, those reflecting a mutation, would often appear very similar to the normal version, risking the immune system targeting the body's normal healthy cells, and potentially side-effects with symptoms matching those of an autoimmune disease. It's something to watch out for, but thankfully one trial published in 2024 indicated this doesn't happen, or at least not disproportionately.[39]

Another challenge is that cancer cells tend to have their own way of shutting down or avoiding an immune attack. As we've seen, many cancer cells display protein molecules at their surface to trigger the brake signals in immune cells, which stop a proper immune reaction from developing. To counter this, personalised cancer vaccines could perhaps be used in combination with checkpoint inhibitors – the medicine we discussed earlier which take the brakes off immune cells to boost their effectiveness. A multi-pronged attack from a combination of treatments is an increasingly used strategy, particularly so for medicines that work with our endlessly complex immune system, and perhaps inevitably for an effective attack on something like cancer.

In spite of such ongoing challenges, prototype cancer vaccines have revealed an interesting phenomenon that could greatly enhance their efficacy. In a small trial where patients were given a personalised vaccine in combination with checkpoint inhibitors to unleash a strong immune response against their cancer, something happened beyond the anticipated reaction from the vaccine itself.[40] As immune cells started killing cancer cells, other protein molecules in the cancer cells were also targeted by the immune system, even though they were not included in the original vaccine. This is astonishing, as it means that a vaccine designed to target one facet of a person's cancer may allow an immune response to develop against all sorts of other aspects too, which should considerably increase the potency of the immune attack and make it much harder, presumably, for any cancer cells to escape. It's probably happening because dying cancer cells trigger additional specific immune responses during the course of the treatment – a phenomenon called 'epitope spreading', epitope being the scientific term for the exact target of an immune response. Epitope spreading could very well improve the chances of success for future cancer vaccines.

Once again a profound sense of wonder arises from the intricacies of how the immune system works, in turn triggering ideas and optimism about new medicines that might help us stay healthy. But epitope spreading does not always work in our favour. It can also occur in the context of an autoimmune disease, which might begin with an unwanted but focused immune response, whose diversification over time makes the problem worse.[41] In such a case we would want to find ways of stopping epitope spreading. Understanding the science better – how exactly does epitope spreading happen in the first place? – might enable us to stimulate the process to improve vaccines and stop it happening in autoimmune disease.

One thing is certain: from all the knowledge we have about immune health, and all the current scientific research to understand the molecular details of the immune system, all sorts of new vaccines, and other medicines and treatments to work alongside the immune system, are on the way.

What other medicines or treatments support immune health?

When you think of the wonderful molecules of human biology, what springs to mind? DNA, perhaps? RNA, neurotransmitters? Living in the shadow of Covid-19 had led many of us to better appreciate one molecule in particular: the antibody. In fact, I think it is no exaggeration to say that the future of humanity depends on us understanding how this molecule works, and how it maintains health – not least because we will need to by the time another pandemic arises, antibodies being such vital components of our immune system, boosted by vaccination.

Antibodies are Y-shaped proteins produced naturally in response to an infection, and come in myriad guises, each exquisitely tailored to facilitate the destruction of their target pathogens with John Wick-like precision. Produced naturally by the body's B cells, they work in many ways, including sticking to germs directly, stopping them from entering human cells, or locking onto cells already infected to tag them for destruction by other components of the immune system. Crucially, not only are antibodies a vital part of our natural defences, but they are also the basis of many of the world's most important medicines, from those used to treat autoimmune disorders like multiple sclerosis to cancer immunotherapies. In fact, a huge segment of the pharmaceutical industry is centred on antibodies, worth over $100 billion a year. And we have not exhausted the potential of these molecules for boosting immune health – far from it.

These days we can transform the basic Y-shaped structure of antibodies like never before.[42] We've been able to manipulate this structure for some years now through genetic engineering or by separating and re-combining parts of the protein, but the tools have reached a level of sophistication that has encouraged a swell of creativity. Let's consider how this can be harnessed to create a new kind of medicine to tackle cancer.

Antibodies can be engineered so that all three 'ends' of the Y-shape are utilised to stick to something different. In such fashion an antibody could bind to two protein molecules on a cancer cell, while another part could trigger an immune cell to attack. Alternatively, an antibody can be engineered to have one part sticking to a cancer cell, another part locking onto a receptor that activates killer T cells, and the third arm could target another T cell protein which boosts the attack, or promotes long-lasting activity.[43] By starting an immune attack against a person's cancer it's hoped that this so-called tri-specific antibody, or immune cell engager, could help those who haven't responded to other immune therapies. For instance, if a person's cancer has evolved to show nothing at its surface that an immune cell can detect on it's own as a signature of cancer, this new medicine could make the immune cell latch onto the cancerous cell and attack.

Other kinds of immune cells can be triggered in this way too. For example, a different tri-specific antibody can lock onto two receptors on NK cells plus one protein molecule on the surface of cancer cells.[44] The specific proteins targeted are two which give a potent signal for the NK cell to kill, especially when triggered together. Of course, any of these might be used in combination too. So complex is the immune system that a double-pronged nudge to get a good attack going on a person's cancer might be better than one. In 2024 something resembling a basic antibody but far more complex – essentially a 4-pronged molecule – was used to harness NK cells against a type of cancer called B cell non-Hodgkin lymphoma.[45] This worked to trigger a potent anti-cancer response in mice and non-human primates, with human trials ongoing.

New treatments for cancer always merit our attention, but the scope of antibody research is greater than cancer treatment, with a far-reaching potential for boosting our defence against infectious diseases. For example, the French company Sanofi has engineered an antibody which locks onto three different parts of a protein molecule on the outside coating of the HIV virus, the thinking being that it would be harder for the virus to mutate and avoid being targeted by something which locks onto three things at once. Results are striking. In a lab dish, one tri-specific antibody could already neutralise 204 out of 208 (98 per cent) of the different versions of HIV. When this designer antibody was given to non-human primates they were protected from the simian form of the virus.[46]

In this case, the tri-specific antibody was designed to target the virus directly. Another approach is to use a tri-specific antibody to force immune cells to latch onto virus-infected cells. In this case the antibody has been engineered to have one part sticking to an HIV-infected cell, with two other parts latching onto killer T cells, making them attack. Again the results look promising in non-human primates, even to the extent of triggering an immune attack on cells harbouring the latent, dormant form of HIV.[47] Targeting that hard-to-get-at dormant form of HIV could really help patients in the future.

Another important development comes from shrinking the size of these molecules. The antibodies typically used in medicine today are quite large, as proteins go, which means they can't always reach their targets easily within tightly packed cells.[48] That's because antibodies – those our body naturally makes, and those used as medicines, based on human antibodies – are inherently large. But we know that antibodies don't *have* to be large molecules because, quite amazingly, antibodies made by llamas and camels are much smaller. These smaller versions of antibodies are called nanobodies. One synthetic nanobody called caplacizumab has already been approved for treating a rare blood disorder called acquired thrombotic thrombocytopenic purpura, or TTP, in which blood platelets form small clots where they shouldn't,[49] sometimes leading to a major health problem. The nanobody locks onto a protein important for platelets sticking together to stop aberrant clotting, thereby preventing strokes or organ failure.

Elsewhere nanobodies have been produced to target snake venom toxin, parasitic worms or the spike protein of coronavirus. They have also been designed to enter cells and stabilise the proteins that would otherwise be destroyed in cystic fibrosis.[50] Their ability to penetrate deep into a tumour is seeing nanobodies being developed as diagnostic tools to help oncologists determine the best course of treatment for a patient's specific cancer type. All this remains experimental, but there's clearly a groundswell of activity around these smaller nanobodies. At some point in the future, nanobodies will surely be the basis of several new medicines.

Other hot ideas in development include CAR T cell therapy, which involves isolating a patient's own T cells and genetically modifying them to include a receptor protein able to target the cancer before the cells are then infused back into the patient. The receptor protein itself is based on antibody structures which recognise a person's own cancer cells. This has been approved in limited circumstances, for example for the treatment of children with acute lymphoblastic leukaemia,[51] but in the future this therapy could be extended dramatically to other types of illness. Immune cells causing problems could be targeted for destruction to control allergies or autoimmunity.

Another theory, at this stage still just an idea, is that B cells could be harvested from a person, engineered using CRISPR gene-editing technology to express a particular antibody, and then infused back into the bloodstream. This could conceivably offer the ability to make an antibody against any specific pathogen, and do away with the need for multiple doses of antibody-based medicines. Maybe a library of B cells could be infused with the capacity to produce a suite of bespoke antibodies to target different versions of any given virus. Out of this explosion in the development of new medicines that work with our immune system, some will potentially be game-changing.

The magic and wonder

When I was young I studied physics first, so I might understand something of the fundamental laws and principles that govern how the universe works: forces, fields, the nature of atoms and time. But as I got older I changed my mind. I switched to studying our immune system, because it felt like one of the most vital frontiers of science for our time, for health, and for understanding who we are and how the human body works. Over the last twenty-five years some of the research done by my lab team has helped advance our knowledge which has led to new ideas for medicines. But beyond all this, for me it's been the magic and wonder held within our infinitely intricate immune system – the scientific poetry of it all – that has been so uplifting, emboldening and enriching. I hope some of this magic and wonder has inspired you too. And I hope everything we've discussed will also help you make more informed choices in your everyday life for living healthier, happier and stronger – and see everything about immune health with a greater wisdom.

Conclusion: the overarching journey

In his well-known book *Thinking, Fast and Slow*, the Nobel Prize winner Daniel Kahneman suggested that there are two brain systems involved in how we think:[1] one for thinking fast, automatically and instinctively; the other for thinking more slowly, consciously and with more awareness. Kahneman used the example of the following simple puzzle. Don't try to solve it; just listen to your intuition.

> *A bat and a ball cost £1.10.*
> *The bat costs £1 more than the ball.*
> *How much does the ball cost?*

Quickly, intuitively, it seems that the ball costs 10p. But this isn't right. Thinking through the puzzle more slowly you realise that if the ball cost 10p and the bat cost £1 more, the bat would cost £1.10, and then the total would be £1.20. So in fact the ball must cost 5p, and then the bat would cost £1.05, so that together they add up to £1.10. The question has only one correct answer, and an initial hunch throws up the wrong solution. The point is that thinking fast is great in lots of situations, but sometimes thinking more slowly works better. As Kahneman said, 'Many people are overconfident, prone to place too much faith in their intuitions.'

It's easy to accept a plausible, superficial or immediately attractive answer without thinking deeply about it. But understanding immune health absolutely requires us to think slowly and deeply. Otherwise we succumb to soundbites, slogans and default sets of opinions and ideas we've either grown up with or that have permeated mainstream culture without clear evidence. Vitamin C is an obvious example. I'd always thought it best to drink orange juice whenever I had a cold, but now, having carefully examined the evidence, I realise this is merely something I was brought up with, which threads all the way back to an evangelical approach to vitamin C by the two-time Nobel winner Linus Pauling over fifty years ago.

Whenever you hear something about immune health which sounds too simple, it probably is. You might hear of a new treatment for a condition someone close to you has, and immediately, almost instinctively, you seize on the hope. Dig into the details, however, and maybe a mouse was cured in a very specific circumstance, or a certain group of people were indeed helped, but later relapsed, or there was some correlation between following a particular activity and having a low risk for a particular illness, but cause and effect hasn't been established and any number of socio-economic factors could actually account for it.

We're all susceptible to the allure of new ideas and easy solutions – just look at the story of Linus Pauling and vitamin C. But everything about immune health requires reflection and critical thinking, which in turn demand time and energy and a willingness to embrace complexity and nuance, things we all need help with. It's one of the reasons I wrote this book: not to offer ringing declarations – even if that's what we really want from health advice – but to give you the tools and background knowledge to navigate this complicated yet inspiring and important realm of science and make good,

CONCLUSION: THE OVERARCHING JOURNEY

informed decisions. It's so difficult to ascertain whether this or that lifestyle choice affects your immune health, and why and to what extent, and the evidence is probably not neat and simple but laced with caveats.

We're accustomed to thinking of science as being precise, and in many ways it is, otherwise we would not have space rockets, microwave ovens or mobile phones that are actually powerful mini-computers. But that doesn't mean science always has clear-cut answers, especially in respect of human health and behaviours, and all the more so when it comes to the immune system and our immune health. Immunology is not imprecise – it is progressing all the time, and scientists are motivated to get answers – but the complexity of the human body is only revealed part by part, theory by theory, and experiment by experiment, and multiple factors are always in play. A holistic view is particularly challenging.

In setting out to understand one thing, immune health, it turns out that we must consider all sorts of other things as well: genes, stress, exercise, weight, nutrition, the microbiome, sleep, mental health, what happens as we age, and how immune-based medicines work. Each is its own scientific frontier. We have discovered all sorts of truths, messy as they may be, and busted a few myths along the way. One overarching truism is that everyone's immune system is wired slightly differently – not better or worse, just differently. Which means there's no 'one-size-fits-all' way of improving immune health. Everything, from stress to ageing, will affect some people's immune health more than others. Having carefully weighed up all the evidence, however, and leaving space for subtlety and difference, we can say that:

1. There is good evidence that vitamins A and D are important for immune health. But the idea that a high dose of vitamin C helps stop a cold, or boosts immune health, has little merit. An evidence-based consensus is a better guide than individual opinions.
2. Gut microbes affect immune health. There's also good evidence that probiotics can impact a person's microbiome. But it is extremely hard to test whether probiotics help protect against disease. Research is fast-moving, however, and the future will see us learn to harness the microbiome more effectively.
3. Keeping to a healthy weight is important for many reasons. Immune cells directly reside within body fat; one reason why inflammation is closely linked to obesity. Fat also produces lots of molecules which affect the immune system generally.
4. The effects of exercise on immune health are complicated and depend on the level of exertion. There are different, even opposing, short- and long-term consequences. Moderate exercise seems beneficial: immune cell numbers increase in number in the blood after exercise. Extreme levels of exercise may weaken immune health in the short term but can be beneficial later in life.
5. There is abundant evidence that stress affects immune health. Stress increases cortisol levels, which tends to dampen immune responses. Stress is part of life and not entirely avoidable, but managing or limiting chronic long-term stress is important.

CONCLUSION: THE OVERARCHING JOURNEY

6. Sleep and immune health are deeply connected. A good night's sleep helps immune health, and vice versa: good immune health helps us sleep well. Much more needs to be understood about each person's individual need for sleep.
7. Immune health and mental health are linked. Inflammation, and blood cytokine levels, seem to directly affect the human brain. If mental health issues arise from an over-active immune system, as in some cases of comorbid depression, and possibly other cases too, some specific types of anti-inflammatory medicine may help.
8. There is no question that ageing impacts immune health in many ways, partly because the body has already spent decades battling germs, and accumulated aged cells which take on new characteristics. A low-level background inflammation seems to be a key problem. We may soon be able to formulate specific treatments and medicines for the elderly.
9. Further understanding of the immune system offers the prospect of personalised vaccines and new human-made versions of re-engineered antibodies.

The breadth of scientific research in immune health nowadays is staggering. It is surely one of the greatest frontiers of science in the twenty-first century. But there's a crucial difference between thinking slowly through all this and Daniel Kahneman's question about the bat and ball. Many questions about immune health do not – at least yet – have a single precise and correct answer. Even if every experiment or analysis we have looked at in this book is statistically valid, peer-reviewed and well-controlled, none can ever tell the whole story.

Throughout the journey we've come on, something else emerges. The contemplation itself becomes its own reward. For me, and I hope you too, thinking about immune health involves thinking about living carefully, deeply, enjoyably, knowingly, even soulfully. Scientific advancements about ourselves have a knack of prompting us to take stock of our place in the world, put things in context and make better decisions.

Let me give an example we haven't so far come across. In an experiment first done in the 1950s, people were given goggles to wear which contained mirrors to turn everything upside-down. Amazingly people adapted, and it wasn't long before they could perform tasks normally and eventually declare that things just seemed the right way up. The instant response is, wow, that's cool! But if we think through the implications more slowly, we see so much more. Such a simple experiment, but how provocative in its revelations of how we 'see', how the brain works, and what reality is. Just by being aware of the extent to which reality is constructed by our brain affects our everyday lives not in a practical way, not by telling us to do something more or less, but profoundly all the same.

Throughout this book we have come across all sorts of experiments that have the potential to shift how we think about ourselves. Knowing that cytokine levels might directly influence mental health may change how some people think about their mental health. That stress leads to an increase in cortisol levels, which dampens immune responses, might lead some people to avoid, or manage, chronic long-term stress more determinedly. Knowing how individualised our

CONCLUSION: THE OVERARCHING JOURNEY

immune systems are may help you understand why something that works for someone you know may not work well for you. All of this makes us more aware of our underlying nature. The science of immune health is, therefore, also about self-reflection and the solace, possibly even happiness, that comes from thinking deeply and slowly through something that is incredibly complicated: you.

Another Nobel Prize winner, the Japanese scientist Osamu Shimomura, made immense contributions to my own field of research. He was interested in how animals communicate with one another by colour, which led him to focus on how a certain type of jellyfish glows green. This doesn't sound like an obvious path towards a Nobel Prize, but it turned out that his work on jellyfish was revolutionary for human biology and medicine. That's because we can repurpose the green glowing protein molecule which he discovered in jellyfish to label human cells, parts of cells and even other protein molecules, to watch their behaviour under a microscope. His scientific achievements are inspiring to me not least because they show how very basic, seemingly esoteric and curiosity-driven research can lead to large, unexpected outcomes. The green glowing jellyfish protein has been used to study the immune system, cancer and other diseases, something Shimomura could not possibly have imagined when he was out with his family every summer on the San Juan Islands around 90 miles north of Seattle, catching jellyfish.[2]

Something Shimomura once said has stuck in my mind. It is especially important, he said, to learn how to learn. What he meant, I've come to think — or at least what I take from it — is that you have to be aware of how every tiny discovery might have a bearing on something else, and eventually, if you recognise and cultivate the self-awareness of what you're engaged upon, it accrues into a far greater enlightenment. To navigate any field of immune health, you have to think deeply and slowly, from lots of angles, about what we know and what we don't know, and then independently, for yourself, about what to do.

Acknowledgements

Immunology is a vast realm of science. None of it – the journey, the knowledge or its implications – is simple. Every discovery arises from the work of countless students, technicians, postdocs, colleagues, collaborators and any number of half-forgotten conversations. We celebrate individuals but, at another level, every scientific achievement is owed to a community. I apologise to anyone who played a role in the work I discuss here but have not named or cited.

I thank my parents, Marilyn and Gerald Davis, for their enduring support, and to whom this book is dedicated. In 2008, my father was diagnosed with multiple myeloma, and his life expectancy was not great. But almost every intervention and drug he took seemed to work well and he survived for another fifteen years. Of course, this was not always easy. In fact, it was not only the medicines which kept him alive, but also the deep love and dedication from my mother, and my brother Colin.

More directly affecting this book, I thank everyone who has been in my research team and guided my thinking over many years. I thank all the staff and students at Imperial College London and the University of Manchester, who have deeply shaped my science, and life in general. I thank Jack Strominger, whose lab at Harvard University was where I began to study the immune system, and everyone who worked with me in his lab at that time.

Will Hammond, my editor at Bodley Head, is always a source of inspiration, and every meeting with him is uplifting

ACKNOWLEDGEMENTS

to me. Laura Reeves, also an editor at Bodley Head, has had a huge influence on this book, helping with its structure, style and beauty. My US editor, Joe Calamia had a great number of extremely important suggestions which improved the text tremendously. I also thank Connor Brown, at Viking, for shaping this book at the outset, and Graham Coster, who masterfully line-edited the final text. Caroline Hardman, my literary agent at Hardman and Swainson, has been exceptionally helpful in my journey for all four books I have written. I am grateful too for comments on the text from Jonathan Worboys and anonymous reviews commissioned by Joe Calamia. I thank Mia Quibell-Smith and Nick Lilly for their enthusiasm in helping getting my books noticed. Some passages in this book build upon articles I have written for newspapers and magazines, and editors there have helped me too. Of course, I have been inspired by any number of other authors as well. I won't name people here, because it would look like I'm name-dropping, but I will try to thank you in person, at a literary festival or bookshop signing. Above all, I thank my wife Katie, and our children Briony and Jack. Couldn't have done it without you. Or maybe I could have, but it would have been absolutely awful.

Glossary

Antibody: a large Y-shaped protein molecule produced by immune cells called B cells. Antibodies can directly attach to disease-causing germs, such as bacteria and viruses, or can stick to human cells infected with germs, or can lock onto toxins or other molecules. You have 10 billion B cells in your body able to produce 10 billion different antibodies, with the idea that at least one would have the right shape to latch onto something that enters the body and is dangerous. Every B cell also has a version of its own antibody tethered to its surface – the B cell receptor – so the cell can tell when there is something in the body its antibody could lock onto. When a B cell does have the right antibody to lock onto something alien and troublesome, its B cell receptor is triggered, and then that particular B cell multiplies so its useful antibody is produced in bulk, ready to neutralise the intruding germ or stop a toxin from working, and so on.

Antioxidant: neutralises the activity of unstable atoms or molecules, called free radicals, that arise from pollutants, toxins and everyday body processes. Vitamin C is an example.

Autoimmune disease: a disease caused by an unwanted immune response against the body's normal healthy cells or tissues. Multiple sclerosis, type 1 diabetes and rheumatoid arthritis are examples.

Bacteria: single-celled microscopic living organisms. There are millions, maybe even billions, of different types.

Colonies of bacteria live on your skin, in your airways and mouth and of, course, throughout your intestine.

Butyrate: a chemical produced during the process by which many gut bacteria gain energy. Butyrate is one of the chemicals produced by gut bacteria that is very well established to directly affect our immune system. There are other short chain fatty acids which are also important, including acetate and propionate, but butyrate is singled out in this book, as perhaps the most potent in affecting the immune system directly, although they all do.

Cell: your body is made up from about 37 trillion cells – muscle cells, skin cells, kidney cells, neurons, and countless others. Each cell in your body carries the exact same set of genes – your personal version of the 23,000 genes which make up the human genome. These genes instruct cells to make protein molecules, and it's these protein molecules which define a cell's job in the body. For example, red blood cells produce haemoglobin, which binds and releases oxygen, and this gives red blood cells their ability to shuttle oxygen around in the body. What makes one type of cell differ from another is which of the 23,000 genes have been switched on for producing protein molecules. The genes for making haemoglobin, for example, are only switched on in red blood cells. Every other cell has its own particular set of genes switched on to produce protein molecules which do that cell's work, such as enabling muscle contraction, digestion, transmit electrical signals and anything else that cells might do.

Checkpoint inhibitor: a cancer treatment which relies on a fundamental aspect of the immune system: its brakes. The immune system has brakes to come on, for example, when a virus has been eradicated from the body.

But if they come on amid a long battle with cancer, this can work against us. So this type of medicine blocks these brakes from working, helping some patients fight their cancer.

Cortisol: a hormone produced by the adrenal gland. It regulates blood sugar levels, and blood pressure, as well as influencing the immune system. Levels increase when we are under stress, to prepare the body for a 'fight-or-flight' response.

Cytokine: a very broad category of molecules in the body which act like hormones, sending messages between our organs and tissues, but specifically linked to the immune system and inflammation.

Dendritic cell: an immune cell which is especially potent at initiating an immune reaction (I write a whole chapter about these cells in my book *The Beautiful Cure*).

Fungi: a group of organisms that includes yeasts, moulds and mushrooms.

Gene: pieces of DNA, primarily used to instruct a cell to produce a specific type of protein molecule. Every one of us has a very similar genetic inheritance – the 23,000 genes which make up the human genome – but there are variations in the specific set of genes we've each inherited and, surprisingly, the genes which vary the most from one person to the next are genes in our immune system. Specifically, they are our MHC genes (short for their long-winded name, the major histocompatibility genes). These genes enable the immune system to detect the presence of things that are alien to the human body. Importantly, these genes vary so much that the versions you have are close to being literally unique. In fact, if we take three of the genes which produce these protein molecules and call them A, B and C there are, across all humanity, around 7,000 different versions of each of the A, B and C genes. In one database containing information for 18 million people, used to help find suitable transplant donors, only four have MHC genes like my own.

Hormone: chemicals produced by the body to act as messages between our organs and tissues.

Hygiene hypothesis: the idea that the immune system needs some level of exposure to germs to develop properly.

Histamine: a molecule produced by immune cells, known to be very important during an allergic reaction, causing various symptoms.

Innate immune response: the first thing likely to happen when a germ enters the body and begins to cause damage is that our so-called innate immune response is triggered. This happens when the immune system directly detects germs causing damage to the body. For this, many types of cell in the human body (not only immune cells) have receptor protein molecules which act as the immune system's 'eyes' to see germs. These receptor protein molecules are shaped to connect like a jigsaw piece with germs. For example, one type of receptor protein molecule directly connects with another kind of molecule only found on the surface of bacteria. When this happens the immune cell 'knows' that bacteria are present. This innate detection process reports the presence of, say, bacteria or a virus, but doesn't let the body know anything more specific about the threat, such as which type of bacteria has invaded, *E. Coli* or *Salmonella* or, even more precisely, which strain of *E. Coli* is present.

Insulin: a hormone produced in the pancreas which helps maintain normal blood sugar levels.

GLOSSARY

Inflammation: a general term to describe the immune response to anything, including an injury or infection. It can be acute, as in the case of a transient immune response to a cut, causing redness and swelling, or chronic, as in the case of a long-term inflammation linked to obesity or ageing.

Killer T cells: immune cells, called T cells, survey other cells for signs of disease. Microscopic fingers protrude out from the ball-like T cell to scan the surface of another cell that is close by. If a T cell detects something that isn't yours, something alien, then it can attack. Killer T cells can directly kill virus-infected cells, for example, as well as send out secretions to alert other immune cells. There are other types of T cell too. Helper T cells tend to be more specialist in getting other cells to respond, and are less able to kill the diseased cells directly.

Microbiome: the community of microbes, including fungi, bacteria and viruses, which live in a particular environment. For this book that means the human body. Very often, the term microbiome is also used as a shorthand to refer to the communities of bacteria living in the human gut, or intestine, although strictly speaking it is much bigger than that.

Natural Killer (NK) cell: immune cells able to directly kill cancer cells, as well as some types of virus-infected cell. Natural Killer cells detect protein molecules on the surface of cancer cells that mark them out as cancerous, for example. If this happens, a Natural Killer cell will latch onto a cancer cell, flatten up against it, kill it, detach from the debris and then move on to attack another cancer cell.

Regulatory T cell, or T reg: immune cells which specialise in turning off the activity of other immune cells. In other words, T regs have the task of ensuring that an immune response doesn't happen against something which does not warrant it, or that an ongoing immune reaction doesn't overshoot.

Protein molecule: large molecules in the body which do all sorts of vital tasks. Types of protein molecule include enzymes, cytokines and antibodies.

Sepsis: a serious medical condition arising when the immune system over-reacts to an infection, leading to unwanted tissue damage and organ failure. Sometimes sepsis can be fatal.

Tumour necrosis factor (TNF): a protein molecule which acts as a messenger produced by the immune system. It is one of the key mediators of inflammation in the body. Blocking the action of TNF, as in anti-TNF therapy, is an important treatment for many inflammatory conditions, such as rheumatoid arthritis. (I wrote a lot more about anti-TNF therapy in an earlier book, *The Beautiful Cure*.)

Virus: viruses are made from the same kind of stuff as all other life on earth: genetic material, protein molecules and so on. But viruses are so simple and basic that they cannot do much on their own. They can't move on their own, or replicate on their own. To replicate, they need to enter a host cell. When a virus enters one of your body's cells, the genetic material of the virus exploits that cell, forcing it to produce protein molecules, to make more virus.

Vitamin: molecules which are established as being important to human health. There are thirteen vitamins in total. Each has its own role to play in the body. Vitamins are needed by the body in relatively small quantities, but if we get too little of any vitamin, this tends to increase the risk of particular health problems arising.

Further reading

Any one book can only ever tell part of a story. Here are a few other books which present other viewpoints on some of the issues we've discussed.

Bill Bryson, *The Body: A Guide for Occupants* (Doubleday, 2019; Illustrated edition 2022)

Edward Bullmore, *The Inflamed Mind: A radical new approach to depression* (Short Books, 2018)

Matthew Cobb, *The Idea of the Brain: A History* (Profile, 2020)

Daniel M. Davis, *The Compatibility Gene* (Allen Lane, 2013)

Daniel M. Davis, *The Beautiful Cure* (Vintage, 2018)

Daniel M. Davis, *The Secret Body* (Vintage, 2021)

Philipp Dettmer, *Immune: A Journey into the Mysterious System that Keeps You Alive* (Hodder and Stoughton, 2021)

David Epstein, *The Sports Gene: Talent, Practice and the Truth About Success* (Yellow Jersey, 2013)

Giamila Fantuzzi, *Body Messages: The Quest for the Proteins of Cellular Communication* (Harvard University Press, 2016)

Russell Foster, *Life Time: The New Science of the Body Clock, and How It Can Revolutionize Your Sleep and Health* (Penguin Life, 2022)

Sarah Gilbert and Catherine Green, *Vaxxers: A Pioneering Moment in Scientific History* (Hodder & Stoughton, 2021)

Stephen S. Hall, *A Commotion in the Blood: Life, Death, and the Immune System* (Henry Holt, 1998).

Tom Ireland, *The Good Virus: The Mysterious Microbes that Rule Our World, Shape Our Health and Can Save Our Future* (Hodder, 2023)

George Johnson, *The Cancer Chronicles: Unlocking Medicine's Deepest Mystery* (Bodley Head, 2013)

Katalin Karikó, *Breaking Through: My Life in Science* (Bodley Head, 2024)

Paul Klenerman, *The Immune System: A Very Short Introduction* (Oxford University Press, 2017)

Armand Leroi, *Mutants: On the Form, Varieties and Errors of the Human Body* (Viking, 2003)

Daniel Lieberman, *Exercised: The Science of Physical Activity, Rest and Health* (Allen Lane, 2020)

Monty Lyman, *The Immune Mind: The New Science of Health* (Torva, 2024)

Theresa MacPhail, *Allergic: How Our Immune System Reacts to a Changing World* (Allen Lane, 2023)

Siddhartha Mukherjee, *The Emperor of All Maladies: A Biography of Cancer* (Scribner, 2010)

Siddhartha Mukherjee, *The Song of the Cell: An Exploration of Medicine and the New Human* (Bodley Head, 2022)

James Nestor, *Breath: The New Science of a Lost Art* (Penguin Life, 2020)

Venki Ramakrishnan, *Why We Die: The New Science of Ageing and Longevity* (Hodder, 2024)

Shilpa Ravella, *A Silent Fire: The Story of Inflammation, Diet and Disease* (Bodley Head, 2023)

Matt Richtel, *An Elegant Defense: The Extraordinary New Science of the Immune System: A Tale in Four Lives* (Mariner Books, 2019)

Andrew Solomon, *Far from the Tree: Parents, Children and the Search for Identity* (Simon & Schuster, 2012)

Susan Sontag, *Illness as a Metaphor* (Farrar, Straus & Giroux, 1978)

Tim Spector, *Food for Life: The New Science of Eating Well* (Jonathan Cape, 2022)

Tim Spector, *Spoon-Fed: Why Almost Everything We've Been Told About Food Is Wrong* (Vintage, 2020)

FURTHER READING

Andrew Steele, *Ageless: The New Science of Getting Older without Getting Old* (Bloomsbury, 2020)

John Tregoning, *Infectious: Pathogens and How We Fight Them* (Oneworld, 2021)

John Tregoning, *Live Forever? A Curious Scientist's Guide to Wellness, Ageing and Death* (Oneworld, 2025)

John Trowsdale, *What the Body Knows: A Guide to the New Science of Our Immune System* (Yale University Press, 2024)

Chris van Tulleken, *Ultra-Processed People: Why Do We All Eat Stuff That Isn't Food . . . and Why Can't We Stop?* (Cornerstone, 2023)

Matthew Walker, *Why We Sleep: The New Science of Sleep and Dreams* (Allen Lane, 2017)

Ed Yong, *I Contain Multitudes: The Microbes Within Us and a Grander View of Life* (Ecco, 2016)

Carl Zimmer, *A Planet of Viruses* (University of Chicago Press, 2011)

Notes

Introduction: How we differ

1. Jolie, A. My medical choice. *New York Times* (14 May 2013).
2. Sun, B. B., *et al.* Genetic associations of protein-coding variants in human disease. *Nature* 603, 95–102 (2022).
3. Brewerton, D. A. Discovery: HLA and disease. *Current opinion in rheumatology* 15, 369–73 (2003).
4. Sheehan, N. J. The ramifications of HLA-B27. *J R Soc Med* 97, 10–14 (2004).
5. Mentzer, A. J., *et al.* Human leukocyte antigen alleles associate with Covid-19 vaccine immunogenicity and risk of breakthrough infection. *Nat Med* (2022).
6. Cooper, G. S. and Stroehla, B. C. The epidemiology of autoimmune diseases. *Autoimmun Rev* 2, 119–125 (2003).
7. Brodin, P. Immune determinants of Covid-19 disease presentation and severity. *Nat Med* 27, 28–33 (2021).
8. Wilkinson, N. M., Chen, H. C., Lechner, M. G. and Su, M. A. Sex differences in immunity. *Annu Rev Immunol* 40, 75–94 (2022).
9. Lakshmikanth, T., *et al.* Immune system adaptation during gender-affirming testosterone treatment. *Nature* 633, 155–164 (2024).
10. McCarthy, M. M. The immune system of trans men reveals how hormones shape immunity. *Nature* 633, 38–40 (2024).

NOTES

1. Orange Juice and Sunshine – or, do vitamins help immune health?

1. Free radicals don't have enough electrons to be stable, so they tend to take electrons from other molecules, including DNA, proteins and cell membrane components, which can be damaging. Vitamin C, and other antioxidants, give some of their own electrons to free radicals, which stabilises them and alleviates the problem.
2. In more detail, iron from food comes in two forms, heme and nonheme. Heme iron comes from meat and fish, while nonheme iron comes from plants including seeds, nuts and grains. Vitamin C has long been known to enhance the body's ability to absorb so-called nonheme iron from food.
3. Pauling, L. The nature of the chemical bond. Application of results obtained from the quantum mechanics and from a theory of paramagnetic susceptibility to the structure of molecules. *Journal of the American Chemical Society* 53, 1367–1400 (1931).
4. Hager, T. *Force of Nature: The Life of Linus Pauling* (Simon and Schuster, New York, 1995).
5. From an interview with Linus Pauling by John L. Heilbron, 17 March 1964. Quoted in Goertzel, T., and Goertzel, B. *Linus Pauling: A Life in Science and Politics* (Basic Books, New York, 1995).
6. Pielke, R. In retrospect: the social function of science. *Nature* 507, 427–8 (2014).
7. This phrase was used by Pauling in a speech which features in a film broadcast in 1961. Available online here (accessed 24 August 2023): https://www.youtube.com/watch?v=kr41MviPXpg
8. Pauling, L. *No More War!* (Dodd, Mead, New York, 1958).

9. Pauling, L. Evolution and the need for ascorbic acid. *Proc Natl Acad Sci* 67, 1643–8 (1970).
10. Stone, I. *The Healing Factor: Vitamin C Against Disease* (Grosset & Dunlap, 1972).
11. To be clear, health authorities in the US and UK today state that such a high level can cause nausea, diarrhoea or vomiting.
12. Cerullo, G., *et al.* The Long History of Vitamin C: From Prevention of the Common Cold to Potential Aid in the Treatment of Covid-19. *Front Immunol* 11, 574029 (2020).
13. Pauling, L. *Vitamin C and the Common Cold* (W. H. Freeman, San Francisco, 1970).
14. In truth, it's very hard to define a 'natural' treatment. Everything is an intervention to some extent, and almost everything is produced or packaged industrially, as well as being derived from nature at some level.
15. Brody, J. E. Vitamin C sales booming despite skepticism on Pauling cold cure. *New York Times* 44 (5 December 1970).
16. Hemila, H. and Chalker, E. Vitamin C for preventing and treating the common cold. *Cochrane Database Syst Rev* 2013, CD000980 (2013).
17. Cameron, E. and Pauling, L. Ascorbic acid and the glycosaminoglycans. An orthomolecular approach to cancer and other diseases. *Oncology* 27, 181–92 (1973).
18. Hanlon, J. Is vitamin C an effective cancer treatment? *New Scientist*, 30–31 (1 July 1976).
19. Creagan, E. T., *et al.* Failure of high-dose vitamin C (ascorbic acid) therapy to benefit patients with advanced cancer. *New England Journal of Medicine* 301, 687–90 (1979).
20. Moertel, C. G. Vitamin C therapy of advanced cancer. *New England Journal of Medicine* 301, 1399–1399 (1979).
21. Moertel, C. G., *et al.* High-Dose Vitamin C versus Placebo in the Treatment of Patients with Advanced Cancer Who Have

Had No Prior Chemotherapy. *New England Journal of Medicine* 312, 137–41 (1985).
22. Hodi, F. S., *et al.* Improved survival with ipilimumab in patients with metastatic melanoma. *New England Journal of Medicine* 363, 711–23 (2010).
23. Wolchok, J. D., *et al.* Guidelines for the evaluation of immune therapy activity in solid tumors: immune-related response criteria. *Clin Cancer Res* 15, 7412–20 (2009).
24. Hoos, A., *et al.* A clinical development paradigm for cancer vaccines and related biologics. *J Immunother* 30, 1–15 (2007).
25. Chauss, D., *et al.* Autocrine vitamin D signaling switches off pro-inflammatory programs of T(H)1 cells. *Nat Immunol* 23, 62–74 (2022).
26. Mora, J. R., Iwata, M. and von Andrian, U. H. Vitamin effects on the immune system: vitamins A and D take centre stage. *Nature Reviews Immunology* 8, 685–98 (2008).
27. Jolliffe, D. A., *et al.* Vitamin D supplementation to prevent acute respiratory infections: a systematic review and meta-analysis of aggregate data from randomised controlled trials. *Lancet Diabetes Endocrinol* 9, 276–92 (2021).
28. Lips, P. Vitamin D to prevent acute respiratory infections. *Lancet Diabetes Endocrinol* 9, 249–51 (2021).
29. Hernandez, J. L., *et al.* Vitamin D status in hospitalized patients with SARS-CoV-2 infection. *J Clin Endocrinol Metab* 106, e1343–53 (2021).
30. Whipple, T. People hit hardest by Covid-19 more likely to lack vitamin D. *The Times (UK)* (28 October 2022).
31. Slomski, A. Vitamin D supplements don't reduce Covid-19 risk. *JAMA* 328, 1581 (2022).
32. Munger, K. L., *et al.* Vitamin D intake and incidence of multiple sclerosis. *Neurology* 62, 60–65 (2004).

33. Miclea, A., Bagnoud, M., Chan, A. and Hoepner, R. A brief review of the effects of vitamin D on multiple sclerosis. *Front Immunol* 11, 781 (2020).
34. Hahn, J., *et al.* Vitamin D and marine omega 3 fatty acid supplementation and incident autoimmune disease: VITAL randomized controlled trial. *British Medical Journal* 376, e066452 (2022).
35. Karohl, C., *et al.* Heritability and seasonal variability of vitamin D concentrations in male twins. *Am J Clin Nutr* 92, 1393–8 (2010).
36. Zhao, S. S., Mason, A., Gjekmarkaj, E., Yanaoka, H. and Burgess, S. Associations between vitamin D and autoimmune diseases: Mendelian randomization analysis. *Seminars in Arthritis and Rheumatism* 62, 152238 (2023).
37. Ali, A., Khan, H., Bahadar, R., Riaz, A. and Asad, M. H. H. B. The impact of airborne pollution and exposure to solar ultraviolet radiation on skin: mechanistic and physiological insight. *Environmental Science and Pollution Research* 27, 28730–6 (2020).
38. Bhat, G. H., Guldin, S., Khan, M. S., Yasir, M. and Prasad, G. Vitamin D status in Psoriasis: impact and clinical correlations. *BMC Nutrition* 8, 115 (2022).
39. Aibana, O., *et al.* Impact of vitamin A and carotenoids on the risk of tuberculosis progression. *Clinical Infectious Diseases* 65, 900–9 (2017).
40. Huang, Z., Liu, Y., Qi, G., Brand, D. and Zheng, S. G. Role of vitamin A in the immune system. *J Clin Med* 7, 258 (2018).
41. Huiming, Y., Chaomin, W. and Meng, M. Vitamin A for treating measles in children. *Cochrane Database Syst Rev* 2005, CD001479 (2005).
42. Gutierrez, S., Svahn, S. L. and Johansson, M. E. Effects of omega-3 fatty acids on immune cells. *Int J Mol Sci* 20, 5028 (2019).

43. Napier, B. A., *et al.* Western diet regulates immune status and the response to LPS-driven sepsis independent of diet-associated microbiome. *Proc Natl Acad Sci* 116, 3688–94 (2019).
44. Corbin, K. D., *et al.* Host-diet-gut microbiome interactions influence human energy balance: a randomized clinical trial. *Nature Communications* 14, 3161 (2023).
45. Sofi, F., Cesari, F., Abbate, R., Gensini, G. F. and Casini, A. Adherence to Mediterranean diet and health status: meta-analysis. *British Medical Journal* 337, a1344 (2008).
46. Jin, Z., Xiang, C., Cai, Q., Wei, X. and He, J. Alcohol consumption as a preventive factor for developing rheumatoid arthritis: a dose-response meta-analysis of prospective studies. *Annals of the Rheumatic Diseases* 73, 1962–7 (2014).
47. Lu, B., *et al.* Alcohol consumption and markers of inflammation in women with preclinical rheumatoid arthritis. *Arthritis & Rheumatism* 62, 3554–9 (2010).
48. To, K., *et al.* The association between alcohol consumption and osteoarthritis: a meta-analysis and meta-regression of observational studies. *Rheumatology International* 41, 1577–91 (2021).

2. The Aliens Living Inside You – or, does the microbiome affect immune health?

1. Various articles online discuss the relationship between the company Danone and the hugely popular computer game, Fortnite, such as this one, from British Brands Group, an organisation championing UK brands (accessed October 2023): https://www.britishbrandsgroup.org.uk/2023/05/17/actimel-builds-immunity-on-fortnite/

NOTES

2. An advert for the in-game Immunity Station in Fortnite is available online here (accessed 19 October 2023): https://www.youtube.com/watch?v=26Wk9njs63c
3. Lampl, M. Obituary for Professor David Barker. *Annals of Human Biology* 41, 187–90 (2014).
4. Barker, D. J., Morris, J. and Nelson, M. Vegetable consumption and acute appendicitis in 59 areas in England and Wales. *British Medical Journal* 292, 927–30 (1986).
5. Strachan, D. P. Hay fever, hygiene, and household size. *British Medical Journal* 299, 1259–60 (1989).
6. Bach, J. F. The hygiene hypothesis in autoimmunity: the role of pathogens and commensals. *Nature Reviews Immunology* 18, 105–20 (2018).
7. Zhong, Y., Zhang, Y., Wang, Y. and Huang, R. Maternal antibiotic exposure during pregnancy and the risk of allergic diseases in childhood: A meta-analysis. *Pediatr Allergy Immunol* 32, 445–56 (2021).
8. Scudellari, M. News Feature: Cleaning up the hygiene hypothesis. *Proc Natl Acad Sci* 114, 1433–6 (2017).
9. Wolf, J., *et al.* Burden of disease attributable to unsafe drinking water, sanitation, and hygiene in domestic settings: a global analysis for selected adverse health outcomes. *Lancet* 401, 2060–71 (2023).
10. Prüss-Ustün, A., *et al.* Burden of disease from inadequate water, sanitation and hygiene for selected adverse health outcomes: An updated analysis with a focus on low- and middle-income countries. *Int J Hyg Environ Health* 222, 765–77 (2019).
11. Bloomfield, S. F., Stanwell-Smith, R., Crevel, R. W. R. and Pickup, J. Too clean, or not too clean: the Hygiene Hypothesis and home hygiene. *Clinical & Experimental Allergy* 36, 402–25 (2006).

12. von Mutius, E. and Vercelli, D. Farm living: effects on childhood asthma and allergy. *Nature Reviews Immunology* 10, 861–8 (2010).
13. Genuneit, J. Exposure to farming environments in childhood and asthma and wheeze in rural populations: a systematic review with meta-analysis. *Pediatric Allergy and Immunology* 23, 509–18 (2012).
14. Chatila, T. A. Innate Immunity in Asthma. *New England Journal of Medicine* 375, 477–9 (2016).
15. Stein, M.M., *et al*. Innate immunity and asthma risk in Amish and Hutterite farm children. *New England Journal of Medicine* 375, 411–21 (2016).
16. Marques Dos Santos, M., *et al*. Asthma-protective agents in dust from traditional farm environments. *J Allergy Clin Immunol* 152, 610–21 (2023).
17. Illi, S., *et al*. Protection from childhood asthma and allergy in Alpine farm environments – the GABRIEL Advanced Studies. *J Allergy Clin Immunol* 129, 1470–7 e1476 (2012).
18. Perdijk, O., van Splunter, M., Savelkoul, H. F. J., Brugman, S. and van Neerven, R. J. J. Cow's milk and immune function in the respiratory tract: potential mechanisms. *Front Immunol* 9, 143 (2018).
19. van Herwijnen, M. J., *et al*. Comprehensive proteomic analysis of human milk-derived extracellular vesicles unveils a novel functional proteome distinct from other milk components. *Mol Cell Proteomics* 15, 3412–23 (2016).
20. Eduard, W., Douwes, J., Omenaas, E. and Heederik, D. Do farming exposures cause or prevent asthma? Results from a study of adult Norwegian farmers. *Thorax* 59, 381–6 (2004).
21. Sigsgaard, T., *et al*. Respiratory diseases and allergy in farmers working with livestock: a EAACI position paper. *Clin Transl Allergy* 10, 29 (2020).

22. Vatanen, T., et al. Variation in microbiome LPS immunogenicity contributes to autoimmunity in humans. *Cell* 165, 842–53 (2016).
23. Hofer, U. Microbiome: Is LPS the key to the hygiene hypothesis? *Nat Rev Microbiol* 14, 334–5 (2016).
24. Routy, B., et al. Gut microbiome influences efficacy of PD-1-based immunotherapy against epithelial tumors. *Science* 359, 91–97 (2018).
25. Gopalakrishnan, V., et al. Gut microbiome modulates response to anti-PD-1 immunotherapy in melanoma patients. *Science* 359, 97–103 (2018).
26. Lee, K. A., et al. Cross-cohort gut microbiome associations with immune checkpoint inhibitor response in advanced melanoma. *Nat Med* 28, 535–44 (2022).
27. Mann, E. R., Lam, Y. K. and Uhlig, H. H. Short-chain fatty acids: linking diet, the microbiome and immunity. *Nature Reviews Immunology* 24, 577–95 (2024).
28. Li, X., Zhang, S., Guo, G., Han, J. and Yu, J. Gut microbiome in modulating immune checkpoint inhibitors. *eBioMedicine* 82 (2022).
29. Schirmer, M., Garner, A., Vlamakis, H. and Xavier, R. J. Microbial genes and pathways in inflammatory bowel disease. *Nature Reviews Microbiology* 17, 497–511 (2019).
30. Shoubridge, A. P., et al. The gut microbiome and mental health: advances in research and emerging priorities. *Mol Psychiatry* 27, 1908–19 (2022).
31. Walter, J., Armet, A. M., Finlay, B. B. and Shanahan, F. Establishing or exaggerating causality for the gut microbiome: lessons from human microbiota-associated rodents. *Cell* 180, 221–32 (2020).
32. Garrett, W. S., et al. Communicable ulcerative colitis induced by T-bet deficiency in the innate immune system. *Cell* 131, 33–45 (2007).

33. Kennedy, J. M., De Silva, A., Walton, G. E. & Gibson, G. R. A review on the use of prebiotics in ulcerative colitis. *Trends in Microbiology* 32, 507–15 (2024).
34. Berer, K., *et al.* Gut microbiota from multiple sclerosis patients enables spontaneous autoimmune encephalomyelitis in mice. *Proc Natl Acad Sci* 114, 10719–24 (2017).
35. Britton, G. J., *et al.* Microbiotas from humans with inflammatory bowel disease alter the balance of gut Th17 and RORgammat(+) regulatory T cells and exacerbate colitis in mice. *Immunity* 50, 212–24 e214 (2019).
36. Li, Y., *et al.* The gut microbiota regulates autism-like behavior by mediating vitamin B(6) homeostasis in EphB6-deficient mice. *Microbiome* 8, 120 (2020).
37. Li, Y., *et al.* (2020).
38. Rendulic, S., *et al.* A predator unmasked: life cycle of Bdellovibrio bacteriovorus from a genomic perspective. *Science* 303, 689–92 (2004).
39. Coyte, K. Z. and Rakoff-Nahoum, S. Understanding competition and cooperation within the mammalian gut microbiome. *Curr Biol* 29, R538–R544 (2019).
40. Claesson, M. J., *et al.* Composition, variability, and temporal stability of the intestinal microbiota of the elderly. *Proc Natl Acad Sci* 108 Suppl 1, 4586–91 (2011).
41. One way of treating antibiotic resistant *C. diff.* (to be used only in severe cases) is by a faecal transplant, which we will come across later. The procedure has its own dangers, though, because faeces can easily be contaminated with other dangerous strains of bacteria.
42. Cong, X., Henderson, W. A., Graf, J. and McGrath, J. M. Early life experience and gut microbiome: the brain-gut-microbiota signaling system. *Adv Neonatal Care* 15, 314–23 (2015).

43. Koch, M. A., *et al.* Maternal IgG and IgA antibodies dampen mucosal T helper cell responses in early life. *Cell* 165, 827–41 (2016).
44. Backhed, F., *et al.* Dynamics and stabilization of the human gut microbiome during the first year of life. *Cell Host Microbe* 17, 690–703 (2015).
45. Derrien, M., Alvarez, A.-S. and de Vos, W. M. The gut microbiota in the first decade of life. *Trends in Microbiology* 27, 997–1010 (2019).
46. Fulde, M., *et al.* Neonatal selection by Toll-like receptor 5 influences long-term gut microbiota composition. *Nature* 560, 489–93 (2018).
47. Johansson, M. E. and Hansson, G. C. Immunological aspects of intestinal mucus and mucins. *Nature Reviews Immunology* 16, 639–49 (2016).
48. Mowat, A. M. and Agace, W. W. Regional specialization within the intestinal immune system. *Nature Reviews Immunology* 14, 667–85 (2014).
49. Hickey, J. W., *et al.* Organization of the human intestine at single-cell resolution. *Nature* 619, 572–84 (2023).
50. Worby, C. J., Olson, B. S., Dodson, K. W., Earl, A. M. and Hultgren, S. J. Establishing the role of the gut microbiota in susceptibility to recurrent urinary tract infections. *J Clin Invest* 132, e158497 (2022).
51. Butyrate is a short chain fatty acid molecule which is important in immune health. But to be clear, there are other short chain fatty acids which are also important, including acetate and propionate. Butyrate is singled out here, as it is perhaps the most potent in affecting the immune system directly, although they all do.
52. A short chain fatty acid is made up from 1–6 carbon atoms, with hydrogen atoms attached, and one end has a carboxyl group (-COOH).

53. Three studies published in 2013, conducted independently, found that gut microbes secrete short-chain fatty acids which promote the production and activity of a type of T cell called a regulatory T cell, which is involved in dampening or regulating other immune cells. These papers are: Arpaia, N., *et al.* Metabolites produced by commensal bacteria promote peripheral regulatory T-cell generation. *Nature* 504, 451–5 (2013); Atarashi, K., *et al.* Treg induction by a rationally selected mixture of Clostridia strains from the human microbiota. *Nature* 500, 232–6 (2013); Smith, P. M., *et al.* The microbial metabolites, short-chain fatty acids, regulate colonic Treg cell homeostasis. *Science* 341, 569–73 (2013).
54. Trompette, A., *et al.* Gut microbiota metabolism of dietary fiber influences allergic airway disease and hematopoiesis. *Nat Med* 20, 159–66 (2014).
55. Takahashi, D., *et al.* Microbiota-derived butyrate limits the autoimmune response by promoting the differentiation of follicular regulatory T cells. *eBioMedicine* 58, 102913 (2020).
56. De Filippis, F., *et al.* Specific gut microbiome signatures and the associated pro-inflammatory functions are linked to pediatric allergy and acquisition of immune tolerance. *Nature Communications* 12, 5958 (2021).
57. Bottcher, M. F., Nordin, E. K., Sandin, A., Midtvedt, T. and Bjorksten, B. Microflora-associated characteristics in faeces from allergic and nonallergic infants. *Clin Exp Allergy* 30, 1590–6 (2000).
58. I discussed this type of immune cell in detail, in my previous book, *The Beautiful Cure* (Vintage, 2019).
59. Gao, Q., *et al.* Bibliometric analysis of global research trends on regulatory T cells in neurological diseases. *Front Neurol* 14, 1284501 (2023).

60. Loverdos, K., et al. Lung microbiome in asthma: current perspectives. *J Clin Med* 8, 1967 (2019).
61. Enaud, R., et al. The gut-lung axis in health and respiratory diseases: a place for inter-organ and inter-kingdom crosstalks. *Front Cell Infect Microbiol* 10, 9 (2020).
62. Klaenhammer, T. R., Kleerebezem, M., Kopp, M.V. and Rescigno, M. The impact of probiotics and prebiotics on the immune system. *Nature Reviews Immunology* 12, 728–34 (2012).
63. Konieczna, P., et al. *Bifidobacterium infantis* 35624 administration induces Foxp3 T regulatory cells in human peripheral blood: potential role for myeloid and plasmacytoid dendritic cells. *Gut* 61, 354–66 (2012).
64. Shadnoush, M., et al. Effects of probiotics on gut microbiota in patients with inflammatory bowel disease: a double-blind, placebo-controlled clinical trial. *Korean J Gastroenterol* 65, 215–21 (2015).
65. Oh, B., et al. The effect of probiotics on gut microbiota during the Helicobacter pylori eradication: randomized controlled trial. *Helicobacter* 21, 165–74 (2016).
66. Beck, L. C., et al. Strain-specific impacts of probiotics are a significant driver of gut microbiome development in very preterm infants. *Nature Microbiology* 7, 1525–35 (2022).
67. Samara, J., et al. Supplementation with a probiotic mixture accelerates gut microbiome maturation and reduces intestinal inflammation in extremely preterm infants. *Cell Host Microbe* 30, 696–711 (2022).
68. Oliphant, K. and Claud, E. C. Early probiotics shape microbiota. *Nature Microbiology* 7, 1506–7 (2022).
69. Bollrath, J. and Powrie, F. Feed your Tregs more fiber. *Science* 341, 463–4 (2013).
70. Bhute, S. S., et al. A high-carbohydrate diet prolongs dysbiosis and *Clostridioides difficile* carriage and increases delayed

mortality in a hamster model of infection. *Microbiology Spectrum* 10, e01804–21 (2022).

71. Hui, W., Li, T., Liu, W., Zhou, C. and Gao, F. Fecal microbiota transplantation for treatment of recurrent *C. difficile* infection: An updated randomized controlled trial meta-analysis. *PLoS One* 14, e0210016 (2019).

72. Strachan, D. P. Rapid response to: 'The hygiene hypothesis' for allergic disease is a misnomer. *British Medical Journal* 349, g5267 (2014).

3. The Evidence of Weight – or, does weight affect immune health?

1. Guidelines issued by the World Obesity Federation, a charity based in London, UK, suggest that we should not call anyone an 'obese person', but rather use the phrasing 'a person with obesity'.

2. NCD Risk Factor Collaboration. Worldwide trends in body-mass index, underweight, overweight, and obesity from 1975 to 2016: a pooled analysis of 2416 population-based measurement studies in 128.9 million children, adolescents, and adults. *Lancet* 390, 2627–42 (2017).

3. World Health Organization, Factsheet on obesity and overweight, published 9 June 2021, accessed May–June 2023, https://www.who.int/news-room/fact-sheets/detail/obesity-and-overweight

4. Ebong, I. A., Goff, D. C., Jr., Rodriguez, C. J., Chen, H. and Bertoni, A. G. Mechanisms of heart failure in obesity. *Obes Res Clin Pract* 8, e540–8 (2014).

5. Hardy, O. T., Czech, M. P. and Corvera, S. What causes the insulin resistance underlying obesity? *Current Opinion in Endocrinology, Diabetes and Obesity* 19, 81–87 (2012).

6. Khan, M. A. B., et al. Epidemiology of Type 2 Diabetes – Global Burden of Disease and Forecasted Trends. *J Epidemiol Glob Health* 10, 107–11 (2020).
7. Matarese, G. The link between obesity and autoimmunity. *Science* 379, 1298–1300 (2023).
8. Hillier, T. A. and Pedula, K. L. Characteristics of an adult population with newly diagnosed type 2 diabetes: the relation of obesity and age of onset. *Diabetes Care* 24, 1522–27 (2001).
9. Watson, T. *Downsizing: How I lost 8 stone, reversed my diabetes and regained my health* (Kyle Books, London, 2020).
10. Taylor, R., Al-Mrabeh, A. and Sattar, N. Understanding the mechanisms of reversal of type 2 diabetes. *Lancet Diabetes & Endocrinology* 7, 726–36 (2019).
11. Lean, M. E., et al. Primary care-led weight management for remission of type 2 diabetes (DiRECT): an open-label, cluster-randomised trial. *Lancet* 391, 541–51 (2018).
12. These follow-up results were announced at the Diabetes UK Professional Conference, Liverpool, on 26 April 2023, and reported in a press release from Newcastle University, published 19 April 2023, here: https://www.ncl.ac.uk/press/articles/latest/2023/04/type2diabetesintoremissionfor5years/
13. Wang, S., et al. Transplantation of chemically induced pluripotent stem-cell-derived islets under abdominal anterior rectus sheath in a type 1 diabetes patient. *Cell* 187, 6152–6164 (2024).
14. Wu, J., et al. Treating a type 2 diabetic patient with impaired pancreatic islet function by personalized endoderm stem cell-derived islet tissue. *Cell Discovery* 10, 45 (2024).
15. Wang, C.-H., et al. CRISPR-engineered human brown-like adipocytes prevent diet-induced obesity and ameliorate metabolic syndrome in mice. *Science Translational Medicine* 12, eaaz8664 (2020).

16. Renehan, A. G., Tyson, M., Egger, M., Heller, R. F. and Zwahlen, M. Body-mass index and incidence of cancer: a systematic review and meta-analysis of prospective observational studies. *Lancet* 371, 569–78 (2008).
17. Onstad, M. A., Schmandt, R. E. & Lu, K. H. Addressing the role of obesity in endometrial cancer risk, prevention, and treatment. *Journal of Clinical Oncology* 34, 4225–30 (2016).
18. Rodriguez, A. C., Blanchard, Z., Maurer, K. A. and Gertz, J. Estrogen signaling in endometrial cancer: a key oncogenic pathway with several open questions. *Hormones and Cancer* 10, 51–63 (2019).
19. Michelet, X., *et al.* Metabolic reprogramming of natural killer cells in obesity limits antitumor responses. *Nat Immunol* 19, 1330–40 (2018).
20. The details here are that mouse Natural Killer cells were isolated from spleen and treated or not treated with fatty acid molecules, palmitate and oleate, for 24 hours, prior to also being bathed in immune-stimulating protein molecules called IL-12, IL-15 and IL-18. The control Natural Killer cells were treated in the same way, just not including the fatty acid molecules. See Michelet, X., *et al.* (2018).
21. Dyck, L., *et al.* Suppressive effects of the obese tumor microenvironment on CD8 T cell infiltration and effector function. *J Exp Med* 219, e20210042 (2022).
22. Lynch, L. A., *et al.* Are natural killer cells protecting the metabolically healthy obese patient? *Obesity* 17, 601–5 (2009).
23. Michelet, X., *et al.* (2018).
24. Orange, J. S. Formation and function of the lytic NK-cell immunological synapse. *Nature Reviews Immunology* (2008).
25. As cancer is rarely caused by a germ, there tends not to be anything as obvious as the presence of molecules from a virus, bacteria or fungi which mark out a cell as being cancerous. In

fact, for a long time it was widely thought that there was nothing at all about a cancerous cell which the immune system could see as a problem: cancer was thought to be hidden from it. However, we now know that protein molecules which have been altered in cancer cells can be detected by immune cells. Also, some types of cancer cell display protein molecules at their surface, signifying that they have become cancerous. Cells have their own in-built ability to know themselves that they have become damaged or become cancerous, and display molecules at their surface signifying this to the immune system.

26. The fact that the immune system has brakes seems obvious now we know they're there. But in truth, the narrative I've just described was only mapped out after brakes on the immune system were stumbled upon in the 1990s by scientists who had set out to understand what switched the immune system on.
27. Tivol, E. A., *et al.* Loss of CTLA-4 leads to massive lymphoproliferation and fatal multiorgan tissue destruction, revealing a critical negative regulatory role of CTLA-4. *Immunity* 3, 541–7 (1995).
28. Waterhouse, P., *et al.* Lymphoproliferative disorders with early lethality in mice deficient in CTLA-4. *Science* 270, 985–8 (1995).
29. Davis, D. M. *The Beautiful Cure* (Vintage, 2018).
30. There are many different receptor protein molecules which can send a switch-off signal to immune cells. Roughly speaking, those present on immune cells constitutively work as checkpoints, to stop an attack on healthy cells. Receptor protein molecules which appear on the surface of immune cells only after the immune system has been in attack mode for some time tend to work to help the system come back down to its normal rest state. Blockade of either type can be beneficial as a therapy. Understanding which receptor would be

best blocked in each person's situation is a vital area of current research.

31. McQuade, J. L., *et al.* Association of body-mass index and outcomes in patients with metastatic melanoma treated with targeted therapy, immunotherapy, or chemotherapy: a retrospective, multicohort analysis. *The Lancet Oncology* 19, 310–22 (2018).

32. Kichenadasse, G., *et al.* Association between body mass index and overall survival with immune checkpoint inhibitor therapy for advanced non-small cell lung cancer. *JAMA Oncology* 6, 512–18 (2020).

33. An, Y., *et al.* Association between body mass index and survival outcomes for cancer patients treated with immune checkpoint inhibitors: a systematic review and meta-analysis. *Journal of Translational Medicine* 18, 235 (2020).

34. Cortellini, A., *et al.* A multicenter study of body mass index in cancer patients treated with anti-PD-1/PD-L1 immune checkpoint inhibitors: when overweight becomes favorable. *Journal for ImmunoTherapy of Cancer* 7, 57 (2019).

35. Boi, S.K., *et al.* Obesity diminishes response to PD-1-based immunotherapies in renal cancer. *Journal for ImmunoTherapy of Cancer* 8, e000725 (2020).

36. Sawadogo, W., Tsegaye, M., Gizaw, A. and Adera, T. Overweight and obesity as risk factors for Covid-19-associated hospitalisations and death: systematic review and meta-analysis. *BMJ Nutrition, Prevention & Health* 5, 10–18 (2022).

37. Simonnet, A., *et al.* High prevalence of obesity in severe acute respiratory syndrome coronavirus-2 (SARS-CoV-2) requiring invasive mechanical ventilation. *Obesity* 28, 1195–9 (2020).

38. Honce, R. and Schultz-Cherry, S. Impact of obesity on influenza A virus pathogenesis, immune response, and evolution. *Front Immunol* 10, 1071 (2019).

39. Almond, M. H., Edwards, M. R., Barclay, W. S. and Johnston, S. L. Obesity and susceptibility to severe outcomes following respiratory viral infection. *Thorax* 68, 684–6 (2013).
40. Dixon, A. E. and Peters, U. The effect of obesity on lung function. *Expert Rev Respir Med* 12, 755–67 (2018).
41. Peralta, G. P., et al. Body mass index and weight change are associated with adult lung function trajectories: the prospective ECRHS study. *Thorax* 75, 313–20 (2020).
42. Winthrop, K. L. and Chiller, T. Preventing and treating biologic-associated opportunistic infections. *Nat Rev Rheumatol* 5, 405–10 (2009).
43. Chen, J., et al. Dose–response relationship between body mass index and tuberculosis in China: a population-based cohort study. *BMJ Open* 12, e050928 (2022).
44. Choi, H., et al. Body mass index, diabetes, and risk of tuberculosis: a retrospective cohort study. *Front Nutr* 8, 739766 (2021).
45. Moreira-Teixeira, L., Mayer-Barber, K., Sher, A. and O'Garra, A. Type I interferons in tuberculosis: foe and occasionally friend. *J Exp Med* 215, 1273–85 (2018).
46. Nasr, M. C., Geerling, E. and Pinto, A. K. Impact of obesity on vaccination to SARS-CoV-2. *Front Endocrinol* 13, 898810 (2022).
47. Painter, S. D., Ovsyannikova, I. G. and Poland, G. A. The weight of obesity on the human immune response to vaccination. *Vaccine* 33, 4422–9 (2015).
48. van der Klaauw, A. A., et al. Accelerated waning of the humoral response to Covid-19 vaccines in obesity. *Nat Med* 29, 1146–54 (2023).
49. Dobner, J. and Kaser, S. Body mass index and the risk of infection – from underweight to obesity. *Clinical Microbiology and Infection* 24, 24–28 (2018).

50. Munteanu, C. and Schwartz, B. The relationship between nutrition and the immune system. *Frontiers in Nutrition* 9, 1082500 (2022).
51. Jordan, S., *et al.* Dietary intake regulates the circulating inflammatory monocyte pool. *Cell* 178, 1102–14.e1117 (2019).
52. Janssen, H., *et al.* Monocytes re-enter the bone marrow during fasting and alter the host response to infection. *Immunity* 56, 783–96.e787 (2023).
53. Domínguez-Andrés, J., Reinecke, H. and Sohrabi, Y. The immune hunger games: the effects of fasting on monocytes. *Cellular & Molecular Immunology* 20, 1098–1100 (2023).
54. Zerwas, S., *et al.* Eating disorders, autoimmune, and autoinflammatory disease. *Pediatrics* 140, e20162089 (2017).
55. Bulik, C. M. and Hardaway, J. A. Turning the tide on obesity? *Science* 381, 463 (2023).
56. Wilding, J. P. H., *et al.* Once-weekly semaglutide in adults with overweight or obesity. *New England Journal of Medicine* 384, 989–1002 (2021).
57. Weghuber, D., *et al.* Once-weekly semaglutide in adolescents with obesity. *New England Journal of Medicine* 387, 2245–57 (2022).

4. *The Yin and Yang of Exercise – or, how much exercise is best?*

1. Paffenbarger, R. S., Jr., Hyde, R. T., Wing, A. L. and Hsieh, C. C. Physical activity, all-cause mortality, and longevity of college alumni. *New England Journal of Medicine* 314, 605–13 (1986).
2. McTiernan, A., *et al.* Physical activity in cancer prevention and survival: A Systematic Review. *Med Sci Sports Exerc* 51, 1252–61 (2019).
3. Pontzer, H., *et al.* Locomotor anatomy and biomechanics of the Dmanisi hominins. *J Hum Evol* 58, 492–504 (2010).

4. van Uffelen, J. G., *et al.* Occupational sitting and health risks: a systematic review. *Am J Prev Med* 39, 379–88 (2010).
5. Lieberman, D. *Exercised: The science of physical activity, rest and health* (Allen Lane, London, 2020).
6. Ahmadi, M. N., Coenen, P., Straker, L. and Stamatakis, E. Device-measured stationary behaviour and cardiovascular and orthostatic circulatory disease incidence. *International Journal of Epidemiology* 53 (2024).
7. Piercy, K. L., *et al.* The physical activity guidelines for Americans. *JAMA* 320, 2020–8 (2018).
8. Warburton, D. E. R. and Bredin, S. S. D. Health benefits of physical activity: a systematic review of current systematic reviews. *Curr Opin Cardiol* 32, 541–56 (2017).
9. Larrabee, R. C. Leucocytosis after violent Exercise. *J Med Res* 7, 76–82 (1902).
10. Walsh, N. P., *et al.* Position statement. Part one: Immune function and exercise. *Exerc Immunol Rev* 17, 6–63 (2011).
11. Shephard, R. J. Adhesion molecules, catecholamines and leucocyte redistribution during and following exercise. *Sports Med* 33, 261–84 (2003).
12. Estaki, M., *et al.* Cardiorespiratory fitness as a predictor of intestinal microbial diversity and distinct metagenomic functions. *Microbiome* 4, 42 (2016).
13. Papayannopoulos, V. Neutrophil extracellular traps in immunity and disease. *Nature Reviews Immunology* 18, 134–47 (2018).
14. Orysiak, J., *et al.* The impact of physical training on neutrophil extracellular traps in young male athletes – a pilot study. *Biol Sport* 38, 459–64 (2021).
15. Lee, J. H. and Jun, H. S. Role of myokines in regulating skeletal muscle mass and function. *Front Physiol* 10, 42 (2019).

16. Jodeiri Farshbaf, M. and Alvina, K. Multiple roles in neuroprotection for the exercise derived myokine irisin. *Front Aging Neurosci* 13, 649929 (2021).
17. Rose-John, S., Winthrop, K. and Calabrese, L. The role of IL-6 in host defence against infections: immunobiology and clinical implications. *Nat Rev Rheumatol* 13, 399–409 (2017).
18. Benatti, F. B. and Pedersen, B. K. Exercise as an anti-inflammatory therapy for rheumatic diseases – myokine regulation. *Nat Rev Rheumatol* 11, 86–97 (2015).
19. Siversten, I. and Dahlstrom, A. W. Relation of muscular activity to carcinoma: a preliminary report. *J. Cancer Res.* 6, 365–78 (1921).
20. Bigley, A. B., *et al.* Acute exercise preferentially redeploys NK-cells with a highly-differentiated phenotype and augments cytotoxicity against lymphoma and multiple myeloma target cells. *Brain Behav Immun* 39, 160–71 (2014).
21. Idorn, M. and Hojman, P. Exercise-dependent regulation of NK cells in cancer protection. *Trends Mol Med* 22, 565–77 (2016).
22. Pedersen, L., *et al.* Voluntary running suppresses tumor growth through epinephrine- and IL-6-dependent NK cell mobilization and redistribution. *Cell Metab* 23, 554–62 (2016).
23. Santiago Zelenay and his team at the Cancer Research UK Manchester Institute, among many others, are working to understand this crucial issue: what determines whether an immune response attacks a tumour, which would be good for the patient, or helps it grow, obviously bad for the patient. They have found out that outcomes depend on what happens very early in a tumour's development. Natural Killer cells often infiltrate a tumour early on. If these immune cells are activated appropriately, they can summon other immune cells into the tumour, and eventually the tumour is killed. But if the tumour

secretes molecules which suppress this early invasion of Natural Killer cells, subsequent waves of the immune response don't happen and the tumour prospers. In the near future, a genetic test might help predict, from a biopsy, what is happening in a person's tumour and this might be used to indicate the likelihood that a cancer patient would benefit from immune therapy, or it would be better to use a different type of therapy.

24. Bonavita, E., *et al.* Antagonistic inflammatory phenotypes dictate tumor fate and response to immune checkpoint blockade. *Immunity* 53, 1215–29 (2020).

25. Bottcher, J. P., *et al.* NK Cells stimulate recruitment of cDC1 into the tumor microenvironment promoting cancer immune control. *Cell* 172, 1022–37 (2018).

26. Rundqvist, H., *et al.* Cytotoxic T-cells mediate exercise-induced reductions in tumor growth. *eLife* 9, e59996 (2020).

27. The way this is done is wondrous in itself, by getting a part of the mouse immune system to attack otherwise healthy cells and remove them. In more detail, an antibody is given to the mouse which locks onto the immune cells which are being removed, in this case the killer T cells. In effect, this tags the cells as if they are diseased, and then the immune system sees them as cells which should be killed off. Of course, a healthy mouse, or any animal, won't naturally make antibodies which target its own healthy cells. But, any animal can make antibodies against a protein molecule found in any other animal. So, for example, a rat can be immunised with a mouse protein molecule to produce antibodies against it. In this particular experiment, the team used a rat-made antibody which targeted mouse killer T cells, tagging them for destruction.

28. In detail, however, this was done using a melanoma cancer cell engineered to be specifically seen by the T cells. So this result serves to highlight that exercise can help T cells work

better at detecting and killing cancer cells, but doesn't immediately mean that T cells from running mice would more effectively deal with the natural development of a tumour.

29. Kurz, E., *et al.* Exercise-induced engagement of the IL-15/IL-15Ralpha axis promotes anti-tumor immunity in pancreatic cancer. *Cancer Cell* 40, 720–37 (2022).
30. Ngo-Huang, A., *et al.* Home-based exercise during preoperative therapy for pancreatic cancer. *Langenbecks Arch Surg* 402, 1175–85 (2017).
31. Fekety, R., *et al.* Epidemiology of antibiotic-associated colitis; isolation of *Clostridium difficile* from the hospital environment. *Am J Med* 70, 906–8 (1981).
32. Pape, K., *et al.* Leisure-time physical activity and the risk of suspected bacterial infections. *Med Sci Sports Exerc* 48, 1737–44 (2016).
33. Nieman, D. C. & Wentz, L. M. The compelling link between physical activity and the body's defense system. *J Sport Health Sci* 8, 201–17 (2019).
34. Zhou, G., *et al.* Smoking, leisure-time exercise and frequency of self-reported common cold among the general population in northeastern China: a cross-sectional study. *BMC Public Health* 18, 294 (2018).
35. Hamer, M., O'Donovan, G. and Stamatakis, E. Lifestyle risk factors, obesity and infectious disease mortality in the general population: Linkage study of 97,844 adults from England and Scotland. *Prev Med* 123, 65–70 (2019).
36. Barrett, B., *et al.* Meditation or exercise for preventing acute respiratory infection: a randomized controlled trial. *Ann Fam Med* 10, 337–46 (2012).
37. Barrett, B., *et al.* Meditation or exercise for preventing acute respiratory infection (MEPARI-2): A randomized controlled trial. *PLoS One* 13, e0197778 (2018).

38. Sim, Y. J., Yu, S., Yoon, K. J., Loiacono, C. M. and Kohut, M. L. Chronic exercise reduces illness severity, decreases viral load, and results in greater anti-inflammatory effects than acute exercise during influenza infection. *J Infect Dis* 200, 1434–42 (2009).
39. Note that the name chickenpox is unfair to chickens, because there's no way the virus can be caught from them. They simply never have this virus. Nobody really knows why it's called chickenpox, but it's often said that it's because someone once thought the red spots typically found on an infected person's skin look like chickpeas.
40. Simpson, R. J., et al. Cardiorespiratory fitness is associated with better control of latent herpesvirus infections in a large ethnically diverse community sample: Evidence from the Texas City Stress and Health Study. *Brain, Behavior, and Immunity* 66, e35 (2017).
41. Calabrese, L. and Neiman, D. C. Exercise, infection and rheumatic diseases: what do we know? *RMD Open* 7, e001644 (2021).
42. Cicchella, A., Stefanelli, C. and Massaro, M. Upper respiratory tract infections in sport and the immune system response. A review. *Biology* 10, 362 (2021).
43. Valtonen, M., et al. Common cold in Team Finland during 2018 Winter Olympic Games (PyeongChang): epidemiology, diagnosis including molecular point-of-care testing (POCT) and treatment. *British Journal of Sports Medicine* 53, 1093–8 (2019).
44. Hoffman, M. D. and Krishnan, E. Health and exercise-related medical issues among 1,212 ultramarathon runners: baseline findings from the Ultrarunners Longitudinal TRAcking (ULTRA) study. *PLoS One* 9, e83867 (2014).
45. de Sousa Fernandes, M. S., et al. Impacts of different triathlon races on systemic cytokine profile and metabolic parameters

in healthy individuals: a systematic review. *BMC Sports Science, Medicine and Rehabilitation* 15, 147 (2023).

46. Peake, J. M., Neubauer, O., Walsh, N. P. and Simpson, R. J. Recovery of the immune system after exercise. *Journal of Applied Physiology* 122, 1077–87 (2017).

47. Campbell, J. P. and Turner, J. E. Debunking the myth of exercise-induced immune suppression: redefining the impact of exercise on immunological health across the lifespan. *Front Immunol* 9, 648 (2018).

48. Dorshkind, K., Montecino-Rodriguez, E. and Signer, R. A. The ageing immune system: is it ever too old to become young again? *Nature Reviews Immunology* 9, 57–62 (2009).

49. Shinkai, S., *et al.* Physical activity and immune senescence in men. *Medicine and Science in Sports and Exercise* 27, 1516–26 (1995).

50. de Araújo, A. L., *et al.* Elderly men with moderate and intense training lifestyle present sustained higher antibody responses to influenza vaccine. *Age* 37, 105 (2015).

51. Minuzzi, L. G., *et al.* Effects of lifelong training on senescence and mobilization of T lymphocytes in response to acute exercise. *Exerc Immunol Rev* 24, 72–84 (2018).

52. Duggal, N. A., Niemiro, G., Harridge, S. D. R., Simpson, R. J. and Lord, J. M. Can physical activity ameliorate immunosenescence and thereby reduce age-related multi-morbidity? *Nature Reviews Immunology* 19, 563–72 (2019).

53. Fiuza-Luces, C., *et al.* The effect of physical exercise on anti-cancer immunity. *Nature Reviews Immunology* 24, 282–93 (2024).

54. Shen, B., *et al.* A mechanosensitive peri-arteriolar niche for osteogenesis and lymphopoiesis. *Nature* 591, 438–44 (2021).

55. Pascoe, A. R., Fiatarone Singh, M. A. and Edwards, K. M. The effects of exercise on vaccination responses: A review of chronic and acute exercise interventions in humans. *Brain, Behavior, and Immunity* 39, 33–41 (2014).

56. Ranadive, S. M., *et al.* Effect of acute aerobic exercise on vaccine efficacy in older adults. *Medicine & Science in Sports & Exercise* 46, 455–461 (2014).
57. Gomes-Santos, I. L., *et al.* Exercise training improves tumor control by increasing CD8(+) T-cell infiltration via CXCR3 signaling and sensitizes breast cancer to immune checkpoint blockade. *Cancer Immunology Research* 9, 765–78 (2021).

5. A Reason for Calm – or, does stress impact immune health?

1. Selye, H. *The stress of my life: a scientist's memoirs* (Van Nostrand Reinhold, New York, 1979).
2. Speakman, J. R. and Keijer, J. Not so hot: optimal housing temperatures for mice to mimic the thermal environment of humans. *Mol Metab* 2, 5–9 (2012).
3. Selye, H. A syndrome produced by diverse nocuous agents. *Nature* 138 (1936).
4. Selye, H. *The Stress of Life, revised edition* (McGraw Hill, New York, 1976).
5. Munteanu, C. and Schwartz, B. The relationship between nutrition and the immune system. *Frontiers in Nutrition* 9, 1082500 (2022).
6. Obituary. Dr. Hans Selye dies in Montreal; studied effects of stress on body. *New York Times* (22 October 1982).
7. Webb, E., *et al.* Assessing individual systemic stress through cortisol analysis of archaeological hair. *Journal of Archaeological Science* 37, 807–12 (2010).
8. Hutmacher, F. Putting stress in historical context: why it is important that being stressed out was not a way to be a person 2,000 years ago. *Front Psychol* 12, 539799 (2021).

9. Breen, M. S., *et al.* Acute psychological stress induces short-term variable immune response. *Brain Behav Immun* 53, 172–82 (2016).
10. Gamble, K. L., Berry, R., Frank, S. J. and Young, M. E. Circadian clock control of endocrine factors. *Nat Rev Endocrinol* 10, 466–75 (2014).
11. Padgett, D. A. and Glaser, R. How stress influences the immune response. *Trends Immunol* 24, 444–8 (2003).
12. Webster, J. I., Tonelli, L. and Sternberg, E. M. Neuroendocrine regulation of immunity. *Annu Rev Immunol* 20, 125–63 (2002).
13. Glaser, R. and Kiecolt-Glaser, J. K. Stress-induced immune dysfunction: implications for health. *Nature Reviews Immunology* 5, 243–51 (2005).
14. Vedhara, K., *et al.* Chronic stress in elderly carers of dementia patients and antibody response to influenza vaccination. *Lancet* 353, 627–31 (1999).
15. Leserman, J., *et al.* Progression to AIDS: the effects of stress, depressive symptoms, and social support. *Psychosom Med* 61, 397–406 (1999).
16. Ortega, V. A., Mercer, E. M., Giesbrecht, G. F. and Arrieta, M. C. Evolutionary significance of the neuroendocrine stress axis on vertebrate immunity and the influence of the microbiome on early-life stress regulation and health outcomes. *Front Microbiol* 12, 634539 (2021).
17. Rodriguez-Galan, M. C., *et al.* Immunocompetence of macrophages in rats exposed to Candida albicans infection and stress. *American Journal of Physiology. Cell Physiology* 284, C111–18 (2003).
18. Konstantinos, A. P. and Sheridan, J. F. Stress and influenza viral infection: modulation of proinflammatory cytokine responses in the lung. *Respir Physiol* 128, 71–77 (2001).

19. Sun, W., et al. Chronic psychological stress impairs germinal center response by repressing miR-155. *Brain Behav Immun* 76, 48–60 (2019).
20. Shou, S., et al. Animal models for COVID-19: hamsters, mouse, ferret, mink, tree shrew, and non-human primates. *Front Microbiol* 12, 626553 (2021).
21. Poller, W. C., et al. Brain motor and fear circuits regulate leukocytes during acute stress. *Nature* 607, 578–84 (2022).
22. Devi, S., et al. Adrenergic regulation of the vasculature impairs leukocyte interstitial migration and suppresses immune responses. *Immunity* 54, 1219–30 (2021).
23. Crucian, B., et al. A case of persistent skin rash and rhinitis with immune system dysregulation onboard the International Space Station. *J Allergy Clin Immunol Pract* 4, 759–62 (2016).
24. Harter, K., et al. Different psychosocial factors are associated with seasonal and perennial allergies in adults: cross-sectional results of the KORA FF4 study. *Int Arch Allergy Immunol* 179, 262–72 (2019).
25. Ayyadurai, S., et al. Frontline science: corticotropin-releasing factor receptor subtype 1 is a critical modulator of mast cell degranulation and stress-induced pathophysiology. *J Leukoc Biol* 102, 1299–1312 (2017).
26. Theoharides, T. C. The impact of psychological stress on mast cells. *Ann Allergy Asthma Immunol* 125, 388–92 (2020).
27. Stojanovich, L. and Marisavljevich, D. Stress as a trigger of autoimmune disease. *Autoimmunity Reviews* 7, 209–13 (2008).
28. Porcelli, B., et al. Association between stressful life events and autoimmune diseases: A systematic review and meta-analysis of retrospective case-control studies. *Autoimmunity Reviews* 15, 325–34 (2016).

29. Bangasser, D. A. and Wicks, B. Sex-specific mechanisms for responding to stress. *Journal of Neuroscience Research* 95, 75–82 (2017).
30. Werbner, M., *et al.* Social-stress-responsive microbiota induces stimulation of self-reactive effector T helper cells. *mSystems* 4, e00292–00218 (2019).
31. Davis, D. M. Dexamethasone: the coronavirus drug 91 years in the making. *Sunday Times* (21 June 2020).
32. Group, R. C., *et al.* Dexamethasone in hospitalized patients with Covid-19. *New England Journal of Medicine* 384, 693–704 (2021).
33. Morgan, D. J. and Davis, D. M. Distinct effects of dexamethasone on human natural killer cell responses dependent on cytokines. *Front Immunol* 8, 432 (2017).
34. Chrousos, G. P. Stress and disorders of the stress system. *Nature Reviews Endocrinology* 5, 374–81 (2009).
35. Ho, R. T., *et al.* The effect of t'ai chi exercise on immunity and infections: a systematic review of controlled trials. *J Altern Complement Med* 19, 389–96 (2013).
36. Wouters, O. J., McKee, M. and Luyten, J. Estimated research and development investment needed to bring a new medicine to market, 2009–18. *JAMA* 323, 844–53 (2020).
37. Openshaw, P. J. M. Using correlates to accelerate vaccinology. *Science* 375, 22–23 (2022).
38. Turakitwanakan, W., Mekseepralard, C. and Busarakumtragul, P. Effects of mindfulness meditation on serum cortisol of medical students. *J Med Assoc Thai* 96 Suppl 1, S90–95 (2013).
39. Hunter, M. R., Gillespie, B. W. and Chen, S. Y. Urban nature experiences reduce stress in the context of daily life based on salivary biomarkers. *Front Psychol* 10, 722 (2019).
40. Jobin, J., Wrosch, C. and Scheier, M. F. Associations between dispositional optimism and diurnal cortisol in a community

sample: When stress is perceived as higher than normal. *Health Psychology* 33, 382–91 (2014).
41. Nicolson, N. A., Peters, M. L. and In den Bosch-Meevissen, Y.M.C. Imagining a positive future reduces cortisol response to awakening and reactivity to acute stress. *Psychoneuroendocrinology* 116, 104677 (2020).
42. Tan, S. Y. and Yip, A. Hans Selye (1907–82): Founder of the stress theory. *Singapore Med J* 59, 170–1 (2018).
43. Petticrew, M. P. and Lee, K. The 'father of stress' meets 'big tobacco': Hans Selye and the tobacco industry. *Am J Public Health* 101, 411–8 (2011).
44. Proctor, R. N. The history of the discovery of the cigarette–lung cancer link: evidentiary traditions, corporate denial, global toll. *Tobacco Control* 21, 87–91 (2012).

6. Inner Beauty Sleep – or, does sleep help immune health?

1. To be accurate here, immune cells are the main producers of cytokines, but a lot of other human cells can produce them too, to some extent, such as liver cells, blood vessel cells and so on. This is an important point in general, that as well as there being professional immune cells, many cells in the body are still capable of some degree of immune defence, mainly because they can detect if they themselves have been invaded by a germ. This is because many cells, not only immune cells, have receptor proteins which can lock onto tell-tale signs of germs, such as components from the outer coating of a virus or bacteria. Many cells can also detect if there is something which is normal for a cell to have, such as genetic material, but it is in a place inside the cell where it shouldn't be, indicating that it is from an invading germ. If

a cell senses any of this, it can produce cytokines to signify that there is a problem.
2. Alcami, A. Viral mimicry of cytokines, chemokines and their receptors. *Nature Reviews Immunology* 3, 36–50 (2003).
3. Patterson, C., Hazime, K. S., Zelenay, S. and Davis, D. M. Prostaglandin E(2) impacts multiple stages of the natural killer cell anti-tumor immune response. *Eur J Immunol* 54, e2350635 (2023).
4. TNF stands for 'tumour necrosis factor', from when it was discovered in 1975 as something released from immune cells which causes tumours to turn black and die. This led to great interest whether TNF could be used to treat cancer patients. But it became clear that the cytokine is quite toxic to the body (even at doses too weak to kill a tumour). TNF's ability, at high doses, to kill tumours turns out to be a bit of a red herring, and just one small thing this molecule can do, but probably not relevant to its actual role in the body. So the name TNF is perhaps not ideal, but history counts for a lot in the names given to biological molecules. As a result, the names of cytokines are complicated to say the least. One family of cytokines, grouped together because they have similarities with one another, consists of interleukin-6 (IL-6), IL-11, IL-27 and IL-31, and then also ciliary neurotrophic factor (CNTF), leukemia inhibitory factor (LIF), cardiotrophin 1 (CT-1), neuropoietin and cardiotrophin-like cytokine (CLC; also known as novel neurotrophin 1 or NNT1), B cell stimulating factor 3 (BSF3), and oncostatin M (OSM).
5. Feldmann, M. Development of anti-TNF therapy for rheumatoid arthritis. *Nature Reviews Immunology* 2, 364–71 (2002).
6. Nilsonne, G., Lekander, M., Akerstedt, T., Axelsson, J. and Ingre, M. Diurnal variation of circulating interleukin-6 in humans: a meta-analysis. *PLoS One* 11, e0165799 (2016).

7. Vgontzas, A. N., *et al.* IL-6 and its circadian secretion in humans. *Neuroimmunomodulation* 12, 131–40 (2005).
8. Zielinski, M. R. and Gibbons, A. J. Neuroinflammation, sleep, and circadian rhythms. *Front Cell Infect Microbiol* 12, 853096 (2022).
9. Imeri, L. and Opp, M. R. How (and why) the immune system makes us sleep. *Nat Rev Neurosci* 10, 199–210 (2009).
10. A nuance here is that some cytokines reach the brain via blood but also, cytokines can be generated within the brain itself by specialist immune cells there. This is discussed more in the next chapter.
11. It is not entirely clear how different cytokines affect each stage in sleep. There is evidence that IL-1, IL-6 and TNF impact slow wave sleep. IL-6 has also been associated with lengthening other sleep stages too.
12. Of note here, mice with an impaired immune system can have problems with learning and memory. This may not relate to sleep, though, as cytokines are probably also important in brain development. See for example, Yirmiya, R. and Goshen, I. Immune modulation of learning, memory, neural plasticity and neurogenesis. *Brain, Behavior, and Immunity* 25, 181–213 (2011).
13. Smith, K. J., Gavey, S., Riddell, N. E., Kontari, P. and Victor, C. The association between loneliness, social isolation and inflammation: A systematic review and meta-analysis. *Neuroscience & Biobehavioral Reviews* 112, 519–41 (2020).
14. Franks, N. P. and Wisden, W. The inescapable drive to sleep: Overlapping mechanisms of sleep and sedation. *Science* 374, 556–9 (2021).
15. Miao, A., *et al.* Brain clearance is reduced during sleep and anesthesia. *Nature Neuroscience* 27, 1046–50 (2024).
16. A textbook case of this is the path of a thin tube inside men called the vas deferens. It transports sperm out from

the testicles to the urethra and is the thin tube which is severed in a vasectomy. Instead of this tube taking a direct path, it follows a route far longer than it needs to, passing over another bit of tubing in the area, the ureter, which connects the kidneys to the bladder. A better design would be for it go under the ureter. But it takes a longer route, probably because the position of the testicles changed as we evolved from our ancestors, and as that move occurred, the vas deferens tube got caught over the ureter rather than going under it.

17. Scheiermann, C., Kunisaki, Y. and Frenette, P. S. Circadian control of the immune system. *Nature Reviews Immunology* 13, 190–8 (2013).
18. Here we're talking about relatively mild, and commonly experienced, levels of sleep deprivation. Other issues can happen if this is more serious. Extreme levels of sleep deprivation can even lead to seizures.
19. Tai, X. Y., Chen, C., Manohar, S. and Husain, M. Impact of sleep duration on executive function and brain structure. *Communications Biology* 5, 201 (2022).
20. Roenneberg, T. *Internal time: chronotypes, social jet lag, and why you're so tired* (Harvard University Press, Cambridge, Massachusetts, 2012).
21. Villanea, F. A. and Schraiber, J. G. Multiple episodes of interbreeding between Neanderthal and modern humans. *Nat Ecol Evol* 3, 39–44 (2019).
22. This is not the same logic as when we discussed whether humans are lazier now than in the past. That's because we don't know for sure the extent to which ancient humans sat or relaxed. But we do know that sunrise and sunset happen at different times, depending on where you are, now and historically, so migration would have certainly exposed humans to this change.

23. Velazquez-Arcelay, K., *et al.* Archaic introgression shaped human circadian traits. *Genome Biology and Evolution* 15, evad203 (2023).
24. Walch, O. J., Cochran, A. and Forger, D. B. A global quantification of 'normal' sleep schedules using smartphone data. *Sci Adv* 2, e1501705 (2016).
25. Roenneberg, T., *et al.* A marker for the end of adolescence. *Curr Biol* 14, R1038–9 (2004).
26. O'Callaghan, F., Muurlink, O. and Reid, N. Effects of caffeine on sleep quality and daytime functioning. *Risk Manag Healthc Policy* 11, 263–71 (2018).
27. Stolicyn, A., *et al.* Comprehensive assessment of sleep duration, insomnia, and brain structure within the UK Biobank cohort. *Sleep* 47, zsad274 (2023).
28. Khan, M. A. and Al-Jahdali, H. The consequences of sleep deprivation on cognitive performance. *Neurosciences Journal* 28, 91–99 (2023).
29. Cobb, M. *The Idea of the Brain: A History* (Profile, London, 2020).
30. Durrington, H. J., Farrow, S. N., Loudon, A. S. and Ray, D. W. The circadian clock and asthma. *Thorax* 69, 90–92 (2014).
31. Sierakowski, S. and Cutolo, M. Morning symptoms in rheumatoid arthritis: a defining characteristic and marker of active disease. *Scand J Rheumatol Suppl* 125, 1–5 (2011).
32. To give an example, cells can change their process for metabolism in low oxygen. There are two major metabolic pathways by which cells get their energy. One is less efficient, but becomes more prominent when oxygen is limited.
33. Phillips, B. G., Wang, Y., Ambati, S., Ma, P. and Meagher, R. B. Airways therapy of obstructive sleep apnea dramatically improves aberrant levels of soluble cytokines involved in autoimmune disease. *Clin Immunol* 221, 108601 (2020).

34. Katz, P., Pedro, S. and Michaud, K. Sleep disorders among individuals with rheumatoid arthritis. *Arthritis Care & Research* 75, 1250–60 (2023).
35. Sang, D., *et al.* Prolonged sleep deprivation induces a cytokine-storm-like syndrome in mammals. *Cell* 186, 5500–5516.e21 (2023).
36. Liu, X., *et al.* Effects of poor sleep on the immune cell landscape as assessed by single-cell analysis. *Communications Biology* 4, 1325 (2021).
37. Åkerstedt, T. Shift work and disturbed sleep/wakefulness. *Occupational Medicine* 53, 89–94 (2003).
38. Kecklund, G. and Axelsson, J. Health consequences of shift work and insufficient sleep. *British Medical Journal* 355, i5210 (2016).
39. Hedstrom, A. K., Akerstedt, T., Olsson, T. and Alfredsson, L. Shift work influences multiple sclerosis risk. *Multiple Sclerosis Journal* 21, 1195–9 (2015).
40. Manouchehri, E., *et al.* Night-shift work duration and breast cancer risk: an updated systematic review and meta-analysis. *BMC Women's Health* 21, 89 (2021).
41. Irwin, M., *et al.* Partial sleep deprivation reduces natural killer cell activity in humans. *Psychosomatic Medicine* 56, 493–8 (1994).
42. Irwin, M., *et al.* Partial night sleep deprivation reduces natural killer and cellular immune responses in humans. *The FASEB Journal* 10, 643–53 (1996).
43. De Lorenzo, Beatriz Helena P., *et al.* Chronic sleep restriction impairs the antitumor immune response in mice. *Neuroimmunomodulation* 25, 59–67 (2018).
44. De Lorenzo, B. H. P., de Oliveira Marchioro, L., Greco, C. R. and Suchecki, D. Sleep-deprivation reduces NK cell number and function mediated by β-adrenergic signalling. *Psychoneuroendocrinology* 57, 134–43 (2015).

45. Chen, Y., *et al.* Sleep duration and the risk of cancer: a systematic review and meta-analysis including dose–response relationship. *BMC Cancer* 18, 1149 (2018).
46. Berisha, A., Shutkind, K. and Borniger, J. C. Sleep disruption and cancer: chicken or the egg? *Front Neurosci* 16, 856235 (2022).
47. Ray, M., Rogers, L. Q., Trammell, R. A. and Toth, L. A. Fatigue and sleep during cancer and chemotherapy: translational rodent models. *Comparative Medicine* 58, 234–45 (2008).
48. Zhou, E. S., Partridge, A. H., Syrjala, K. L., Michaud, A. L. and Recklitis, C. J. Evaluation and treatment of insomnia in adult cancer survivorship programs. *J Cancer Surviv* 11, 74–79 (2017).
49. Slade, A. N., Waters, M. R. and Serrano, N. A. Long-term sleep disturbance and prescription sleep aid use among cancer survivors in the United States. *Supportive Care in Cancer* 28, 551–60 (2020).
50. Spiegel, K., Sheridan, J. F. and Van Cauter, E. Effect of sleep deprivation on response to immunizaton. *JAMA* 288, 1471–2 (2002).
51. Spiegel, K., *et al.* A meta-analysis of the associations between insufficient sleep duration and antibody response to vaccination. *Current Biology* 33, 998–1005.e1002 (2023).
52. Opp, M. R. Sleep: Not getting enough diminishes vaccine responses. *Current Biology* 33, R192–R194 (2023).
53. Engler, R. J. M., *et al.* Half- vs full-dose trivalent inactivated influenza vaccine (2004–5): age, dose, and sex effects on immune responses. *Archives of Internal Medicine* 168, 2405–14 (2008).
54. This calculation is only an approximation. One difficulty here is that the effects of sleep have been studied for vaccines against flu and hepatitus, not for Covid-19 vaccines. The

calculation is reported in Spiegel, K., *et al. Current Biology* 33, 998–1005.e1002 (2023).
55. The Lancet, R. A wake-up call for sleep in rheumatic diseases. *The Lancet Rheumatology* 4, e739 (2022).
56. de Lange, M. A., Richmond, R. C., Eastwood, S. V. and Davies, N. M. Insomnia symptom prevalence in England: a comparison of cross-sectional self-reported data and primary care records in the UK Biobank. *BMJ Open* 14, e080479 (2024).
57. Sabiniewicz, A., Zimmermann, P., Ozturk, G. A., Warr, J. and Hummel, T. Effects of odors on sleep quality in 139 healthy participants. *Sci Rep* 12, 17165 (2022).
58. Perez-Pozuelo, I., *et al.* The future of sleep health: a data-driven revolution in sleep science and medicine. *npj Digital Medicine* 3, 42 (2020).

7. The Two Big Systems – or, does immune health affect mental health?

1. Bottazzo, G. F., Pujol-Borrell, R., Hanafusa, T. and Feldmann, M. Role of aberrant HLA-DR expression and antigen presentation in induction of endocrine autoimmunity. *Lancet* 2, 1115–9 (1983).
2. This is discussed in more detail in Davis, D. M. *The Beautiful Cure* (Vintage, London, 2018).
3. This is discussed in more detail in Vilček, J. *Love and Science: a memoir* (Seven Stories Press, New York, 2016).
4. Feldmann, M. Translating molecular insights in autoimmunity into effective therapy. *Annu Rev Immunol* 27, 1–27 (2009).
5. Marshall, N. J., Wilson, G., Lapworth, K. and Kay, L. J. Patients' perceptions of treatment with anti-TNF therapy

for rheumatoid arthritis: a qualitative study. *Rheumatology* 43, 1034–8 (2004).
6. Bullmore, E. *The Inflamed Mind: A radical new approach to depression* (Short Books, London, 2018).
7. Monaco, C., Nanchahal, J., Taylor, P. and Feldmann, M. Anti-TNF therapy: past, present and future. *International Immunology* 27, 55–62 (2014).
8. Siebenhüner, A. R., *et al.* Effects of anti-TNF therapy and immunomodulators on anxiety and depressive symptoms in patients with inflammatory bowel disease: a 5-year analysis. *Therapeutic Advances in Gastroenterology* 14, 17562848211033763 (2021).
9. Almeida, C., *et al.* Biologic interventions for fatigue in rheumatoid arthritis. *Cochrane Database of Systematic Reviews* (2016).
10. Hess, A., *et al.* Blockade of TNF-α rapidly inhibits pain responses in the central nervous system. *Proceedings of the National Academy of Sciences* 108, 3731–6 (2011).
11. In more detail, in the 1980s, it was generally considered that depression suppresses the immune system; but now we tend to think of suppression of the immune system happening because of stress – which increases cortisol levels and dampens immune activity. Of course, stress and depression are not entirely separate but, roughly speaking, stress arises from emotional or mental pressure connected to specific problems like moving house or getting divorced, while depression is more of a general mood disorder, associated with negative thoughts, worries and feelings of worthlessness. In the 1990s, a few researchers, writing in relatively obscure scientific journals, suggested that immune activity might directly relate to the onset of depression. They suggested this because, all too often, people with depression showed signs of an overly active immune system – the opposite to their immune system being suppressed. For example, they said, patients with an

autoimmune condition like rheumatoid arthritis, which arises from an unwanted excessive immune response, also have an increased likelihood of depression. The anti-TNF story recounted here helped bring this big idea, that immune health and mental health are linked, into the limelight.

12. Lee, C. H. and Giuliani, F. The Role of inflammation in depression and fatigue. *Front Immunol* 10, 1696 (2019).
13. Anderson, R. J., Freedland, K. E., Clouse, R. E. and Lustman, P. J. The prevalence of comorbid depression in adults with diabetes: a meta-analysis. *Diabetes Care* 24, 1069–78 (2001).
14. Matcham, F., Rayner, L., Steer, S. and Hotopf, M. The prevalence of depression in rheumatoid arthritis: a systematic review and meta-analysis. *Rheumatology* 52, 2136–48 (2013).
15. Boeschoten, R. E., *et al.* Prevalence of depression and anxiety in Multiple Sclerosis: A systematic review and meta-analysis. *Journal of the Neurological Sciences* 372, 331–41 (2017).
16. Gold, S. M., *et al.* Comorbid depression in medical diseases. *Nature Reviews Disease Primers* 6, 69 (2020).
17. Kingston, A., *et al.* Projections of multi-morbidity in the older population in England to 2035: estimates from the Population Ageing and Care Simulation (PACSim) model. *Age and Ageing* 47, 374–80 (2018).
18. McGrath, J. J., *et al.* Age of onset and cumulative risk of mental disorders: a cross-national analysis of population surveys from 29 countries. *The Lancet Psychiatry* 10, 668–81 (2023).
19. If one disease affects or causes another, this is referred to as comorbidity. But if there just happens to be more than one condition affecting a person – not one causing the other – this is referred to as multi-morbidity.
20. Luppino, F. S., *et al.* Overweight, obesity, and depression: A systematic review and meta-analysis of longitudinal studies. *Archives of General Psychiatry* 67, 220–9 (2010).

21. Nowakowska, M., *et al.* The comorbidity burden of type 2 diabetes mellitus: patterns, clusters and predictions from a large English primary care cohort. *BMC Medicine* 17, 145 (2019).
22. Du, Y.-J., *et al.* Airway inflammation and hypothalamic-pituitary-adrenal axis activity in asthmatic adults with depression. *Journal of Asthma* 50, 274–81 (2013).
23. Gold, S. M., *et al.* Endocrine and immune substrates of depressive symptoms and fatigue in multiple sclerosis patients with comorbid major depression. *Journal of Neurology, Neurosurgery & Psychiatry* 82, 814–8 (2011).
24. Wong, M., *et al.* TNFα blockade in human diseases: Mechanisms and future directions. *Clinical Immunology* 126, 121–36 (2008).
25. Hoang, H., Laursen, B., Stenager, E. N. and Stenager, E. Psychiatric co-morbidity in multiple sclerosis: The risk of depression and anxiety before and after MS diagnosis. *Multiple Sclerosis Journal* 22, 347–53 (2016).
26. Bernstein, C. N., *et al.* Rising incidence of psychiatric disorders before diagnosis of immune-mediated inflammatory disease. *Epidemiology and Psychiatric Sciences* 28, 333–42 (2019).
27. One explanation for this could be that stress may affect mental health and the likelihood of multiple sclerosis arising. Some people with multiple sclerosis do report that stress aggravates their condition. But there's no evidence that stress caused by divorce, or the loss of a child or a spouse, is a risk factor for developing this autoimmune disease in the first place.
28. Brunner, E. J., *et al.* Associations of C-reactive protein and interleukin-6 with cognitive symptoms of depression: 12-year follow-up of the Whitehall II study. *Psychological Medicine* 39, 413–23 (2009).
29. Bell, J. A., *et al.* Repeated exposure to systemic inflammation and risk of new depressive symptoms among older adults. *Translational Psychiatry* 7, e1208–e1208 (2017).

30. Khandaker, G. M., Pearson, R. M., Zammit, S., Lewis, G. and Jones, P. B. Association of serum interleukin 6 and C-reactive protein in childhood with depression and psychosis in young adult life: a population-based longitudinal study. *JAMA Psychiatry* 71, 1121–8 (2014).
31. Zürcher, S. J., *et al*. Post-viral mental health sequelae in infected persons associated with Covid-19 and previous epidemics and pandemics: Systematic review and meta-analysis of prevalence estimates. *Journal of Infection and Public Health* 15, 599–608 (2022).
32. Köhler, C. A., *et al*. Peripheral cytokine and chemokine alterations in depression: a meta-analysis of 82 studies. *Acta Psychiatrica Scandinavica* 135, 373–87 (2017).
33. Nutt, D., Wilson, S. and Paterson, L. Sleep disorders as core symptoms of depression. *Dialogues in Clinical Neuroscience* 10, 329–36 (2008).
34. Lenczowski, M. J., *et al*. Central administration of rat IL-6 induces HPA activation and fever but not sickness behavior in rats. *Am J Physiol* 276, R652–8 (1999).
35. Dantzer, R. Cytokine, sickness behavior, and depression. *Immunol Allergy Clin North Am* 29, 247–64 (2009).
36. Nie, X., *et al*. The innate immune receptors TLR2/4 mediate repeated social defeat stress-induced social avoidance through prefrontal microglial activation. *Neuron* 99, 464–79 e467 (2018).
37. Filiano, A. J., *et al*. Unexpected role of interferon-γ in regulating neuronal connectivity and social behaviour. *Nature* 535, 425–9 (2016).
38. In this case, an interesting scientific detail is that another cytokine produced by immune cells, called interferon-gamma, seemed to be especially important in regulating behaviour.

39. Kitaoka, S. Inflammation in the brain and periphery found in animal models of depression and its behavioral relevance. *Journal of Pharmacological Sciences* 148, 262–6 (2022).
40. There is no clear evidence, for example, that an animal can die by suicide, from grief, loss or depression. Animals can act to kill themselves, such as when a bee stings and then, as it tries to fly away, loses its innards and dies. But this is not the same as dying by suicide from depression. Here a specific behaviour has evolved, probably because the sacrifice of one bee serves to protect others in the hive.
41. Haroon, E., *et al.* Antidepressant treatment resistance is associated with increased inflammatory markers in patients with major depressive disorder. *Psychoneuroendocrinology* 95, 43–49 (2018).
42. Smith, C. J., *et al.* Probiotics normalize the gut-brain-microbiota axis in immunodeficient mice. *American Journal of Physiology-Gastrointestinal and Liver Physiology* 307, G793–G802 (2014).
43. To be clear, there are other experiments published which don't necessarily fit with this one. For example, an experiment with hamsters concluded that probiotics had the opposite effect, decreasing social interactions, for example, making hamsters more 'anxious'. The details of this experiment are quite different, and didn't involve specifoc manipulations of the immune system, like the one with genetically modified mice. The authors of the study with hamsters suggest that the differences across studies could relate to different doses, species or the precise way in which experiments are performed: Partrick, K. A., *et al.* Ingestion of probiotic (*Lactobacillus helveticus* and *Bifidobacterium longum*) alters intestinal microbial structure and behavioral expression following social defeat stress. *Scientific Reports* 11, 3763 (2021).

44. Valles-Colomer, M., *et al*. The neuroactive potential of the human gut microbiota in quality of life and depression. *Nat Microbiol* 4, 623–32 (2019).
45. Liu, L., *et al*. Gut microbiota and its metabolites in depression: from pathogenesis to treatment. *eBioMedicine* 90, 104527 (2023).
46. Checa-Ros, A., Jeréz-Calero, A., Molina-Carballo, A., Campoy, C. and Muñoz-Hoyos, A. Current evidence on the role of the gut microbiome in ADHD pathophysiology and therapeutic implications. *Nutrients* 13, 249 (2021).
47. Dickerson, F., Severance, E. and Yolken, R. The microbiome, immunity, and schizophrenia and bipolar disorder. *Brain, Behavior, and Immunity* 62, 46–52 (2017).
48. Cryan, J. F. and Dinan, T. G. Mind-altering microorganisms: the impact of the gut microbiota on brain and behaviour. *Nat Rev Neurosci* 13, 701–12 (2012).
49. Makris, A. P., Karianaki, M., Tsamis, K. I. and Paschou, S. A. The role of the gut-brain axis in depression: endocrine, neural, and immune pathways. *Hormones* 20, 1–12 (2021).
50. Pinto-Sanchez, M. I., *et al*. Probiotic *Bifidobacterium longum* NCC3001 reduces depression scores and alters brain activity: A pilot study in patients with irritable bowel syndrome. *Gastroenterology* 153, 448–59.e448 (2017).
51. Purcell, S. M., *et al*. Common polygenic variation contributes to risk of schizophrenia and bipolar disorder. *Nature* 460, 748–52 (2009).
52. Stefansson, H., *et al*. Common variants conferring risk of schizophrenia. *Nature* 460, 744–7 (2009).
53. Burgdorf, K. S., *et al*. Large-scale study of Toxoplasma and Cytomegalovirus shows an association between infection and serious psychiatric disorders. *Brain, Behavior, and Immunity* 79, 152–8 (2019).

54. Ingram, W. M., Goodrich, L. M., Robey, E. A. and Eisen, M. B. Mice infected with low-virulence strains of *Toxoplasma gondii* lose their innate aversion to cat urine, even after extensive parasite clearance. *PLoS One* 8, e75246 (2013).
55. Miller, B. J., Buckley, P., Seabolt, W., Mellor, A. and Kirkpatrick, B. Meta-analysis of cytokine alterations in schizophrenia: clinical status and antipsychotic effects. *Biol Psychiatry* 70, 663–71 (2011).
56. Goldsmith, D. R., Rapaport, M. H. and Miller, B. J. A meta-analysis of blood cytokine network alterations in psychiatric patients: comparisons between schizophrenia, bipolar disorder and depression. *Molecular Psychiatry* 21, 1696–1709 (2016).
57. Bayer, T. A., Buslei, R., Havas, L. and Falkai, P. Evidence for activation of microglia in patients with psychiatric illnesses. *Neuroscience Letters* 271, 126–8 (1999).
58. Bloomfield, P. S., *et al.* Microglial activity in people at ultra high risk of psychosis and in schizophrenia: An [(11)C]PBR28 PET brain imaging study. *American Journal of Psychiatry* 173, 44–52 (2016).
59. Sullivan, P. F., Neale, M. C. and Kendler, K. S. Genetic epidemiology of major depression: review and meta-analysis. *American Journal of Psychiatry* 157, 1552–1562 (2000).
60. Howard, D. M., *et al.* Genome-wide meta-analysis of depression identifies 102 independent variants and highlights the importance of the prefrontal brain regions. *Nature Neuroscience* 22, 343–52 (2019).
61. Tubbs, J. D., Ding, J., Baum, L. and Sham, P. C. Immune dysregulation in depression: Evidence from genome-wide association. *Brain, Behavior, & Immunity – Health* 7, 100108 (2020).
62. Zipp, F., Bittner, S. and Schafer, D. P. Cytokines as emerging regulators of central nervous system synapses. *Immunity* 56, 914–25 (2023).

63. Bourgognon, J.-M. abnd Cavanagh, J. The role of cytokines in modulating learning and memory and brain plasticity. *Brain and Neuroscience Advances* 4, 23982128820979802 (2020).
64. Rudolph, M. D., et al. Maternal IL-6 during pregnancy can be estimated from newborn brain connectivity and predicts future working memory in offspring. *Nature Neuroscience* 21, 765–772 (2018).
65. Lee, B. K., et al. Maternal hospitalization with infection during pregnancy and risk of autism spectrum disorders. *Brain, Behavior, and Immunity* 44, 100–105 (2015).
66. Jaswa, E. G., et al. In utero exposure to maternal COVID-19 vaccination and offspring neurodevelopment at 12 and 18 months. *JAMA Pediatrics* 178, 258–265 (2024).
67. Berk, M., et al. Youth depression alleviation with anti-inflammatory agents (YoDA-A): a randomised clinical trial of rosuvastatin and aspirin. *BMC Medicine* 18, 16 (2020).
68. Berk, M., et al. Effect of aspirin vs placebo on the prevention of depression in older people: A Randomized Clinical Trial. *JAMA Psychiatry* 77, 1012–20 (2020).
69. Raison, C. L., et al. A randomized controlled trial of the tumor necrosis factor antagonist Infliximab for treatment-resistant depression: the role of baseline inflammatory biomarkers. *JAMA Psychiatry* 70, 31–41 (2013).
70. McIntyre, R. S., et al. Efficacy of adjunctive Infliximab vs placebo in the treatment of adults with bipolar I/II depression: A randomized clinical trial. *JAMA Psychiatry* 76, 783–90 (2019).
71. Inamdar, A., et al. Evaluation of antidepressant properties of the p38 MAP kinase inhibitor losmapimod (GW856553) in Major Depressive Disorder: Results from two randomised, placebo-controlled, double-blind, multicentre studies using a Bayesian approach. *Journal of Psychopharmacology* 28, 570–81 (2014).

72. Drevets, W. C., Wittenberg, G. M., Bullmore, E. T. and Manji, H. K. Immune targets for therapeutic development in depression: towards precision medicine. *Nat Rev Drug Discovery* 21, 224–44 (2022).

8. *100-Year-Long Immunity – or, how does immune health change as we age?*

1. Simon, A. K., Hollander, G. A. and McMichael, A. Evolution of the immune system in humans from infancy to old age. *Proc Biol Sci* 282, 20143085 (2015).
2. *Global Health and Aging*, a report from the National Institute on Aging (USA) and World Health Organization, available online here: https://www.nia.nih.gov/research/publication/global-health-and-aging/preface
3. The Office for National Statistics (UK) produces annual population data for the UK, available online here: https://www.ons.gov.uk/peoplepopulationandcommunity/populationandmigration/populationestimates The UK charity Age UK also produce a monthly collection of statistics about elderly people, available online here: http://www.ageuk.org.uk/professional-resources-home/
4. Needless to say, there is a lot of complexity in how evolutionary processes affect how we age. It is not intuitive, and mathematical analyses reveal various possibilities to what happens in detail. One example is here: Giaimo, S. and Traulsen, A. The selection force weakens with age because ageing evolves and not vice versa. *Nature Communications* 13, 686 (2022).
5. Franceschi, C., *et al.* Inflamm-aging. An evolutionary perspective on immunosenescence. *Ann N Y Acad Sci* 908, 244–54 (2000).

6. Ma, L., *et al.* Role of interleukin-6 to differentiate sepsis from non-infectious systemic inflammatory response syndrome. *Cytokine* 88, 126–35 (2016).
7. Maggio, M., Guralnik, J. M., Longo, D. L. and Ferrucci, L. Interleukin-6 in aging and chronic disease: a magnificent pathway. *The Journals of Gerontology: Series A* 61, 575–84 (2006).
8. Rea, I. M., *et al.* Age and age-related diseases: role of inflammation triggers and cytokines. *Front Immunol* 9, 586 (2018).
9. Ershler, W. B. and Keller, E. T. Age-associated increased Interleukin-6 gene expression, late-life diseases, and frailty. *Annual Review of Medicine* 51, 245–70 (2000).
10. Oftentimes, words do not capture what we want to say as precisely as we would like, but this is especially true for scientific issues. The fact is that there is a slightly increased level of cytokines in blood as we age. Putting this into plain English is hard. This is partly due to a fuzziness in our knowledge, but also a problem of language and science. In the same way that it's hard to describe quantum physics or relativity in everyday words, because mathematics describes these things, biology has its own symbols and jargon necessary to communicate the nuances between experts, and it is hard to 'translate' this. So we can say that 'the immune system is slightly aroused', which grasps some essence of what's going on, but also there's a vagueness to these words which is not quite satisfying. In a way, that's part of the wonder of it all – that, like quantum physics and relativity, it's beyond words.
11. Shaw, A. C., Goldstein, D. R. and Montgomery, R. R. Age-dependent dysregulation of innate immunity. *Nature Reviews Immunology* 13, 875–87 (2013).
12. Dorshkind, K., Montecino-Rodriguez, E. and Signer, R. A. The ageing immune system: is it ever too old to become young again? *Nature Reviews Immunology* 9, 57–62 (2009).

13. Verity, R., *et al.* Estimates of the severity of coronavirus disease 2019: a model-based analysis. *Lancet Infect Dis* 20, 669–77 (2020).
14. Szymkowicz, S. M., Gerlach, A. R., Homiack, D. and Taylor, W. D. Biological factors influencing depression in later life: role of aging processes and treatment implications. *Translational Psychiatry* 13, 160 (2023).
15. House of Lords Science and Technology Select Committee. Corrected oral evidence: Ageing: Science, Technology and Healthy Living. 29th October 2019. Available online here: https://committees.parliament.uk/oralevidence/9723/html/
16. Munoz-Espin, D. and Serrano, M. Cellular senescence: from physiology to pathology. *Nat Rev Mol Cell Biol* 15, 482–96 (2014).
17. Franco, A. C., Aveleira, C. and Cavadas, C. Skin senescence: mechanisms and impact on whole-body aging. *Trends Mol Med* 28, 97–109 (2022).
18. To distinguish this from how damage to a cell causes it to become senescent, this process is sometimes called replicative senescence.
19. Schmutz, I., *et al.* TINF2 is a haploinsufficient tumor suppressor that limits telomere length. *eLife* 9, e61235 (2020).
20. Harley, C. B. Telomerase and cancer therapeutics. *Nat Rev Cancer* 8, 167–79 (2008).
21. Vyas, C. M., *et al.* Telomere length and its relationships with lifestyle and behavioural factors: variations by sex and race/ethnicity. *Age and Ageing* 50, 838–46 (2020).
22. Morla, M., *et al.* Telomere shortening in smokers with and without COPD. *Eur Respir J* 27, 525–8 (2006).
23. Valdes, A. M., *et al.* Obesity, cigarette smoking, and telomere length in women. *Lancet* 366, 662–4 (2005).
24. There are plenty of other modifications to the human body which happen on a molecular scale as we age, such as changes in glycosylation of some proteins, but these are less well

understood in ageing, and probably less easily targeted, compared to telomere length.

25. Zhang, Q. An interpretable biological age. *Lancet Healthy Longev* 4, e662–e663 (2023).

26. Lasry, A. and Ben-Neriah, Y. Senescence-associated inflammatory responses: aging and cancer perspectives. *Trends in Immunology* 36, 217–28 (2015).

27. Kale, A., Sharma, A., Stolzing, A., Desprez, P. Y. and Campisi, J. Role of immune cells in the removal of deleterious senescent cells. *Immun Ageing* 17, 16 (2020).

28. Evidence for this is that some viruses produce their own protein molecules to stop their host cells becoming senescent. However, there's a slip side to this too. Some viruses have developed strategies so that they do even better when a cell turns senescent. For example, there is evidence that HIV benefits from host cells becoming senescent.

29. Seoane, R., Vidal, S., Bouzaher, Y. H., El Motiam, A. and Rivas, C. The interaction of viruses with the cellular senescence response. *Biology* 9, 455 (2020).

30. Kumari, R. and Jat, P. Mechanisms of cellular senescence: cell cycle arrest and senescence associated secretory phenotype. *Front Cell Dev Biol* 9, 645593 (2021).

31. Rodriguez, I. J., *et al.* Immunosenescence study of T Cells: a systematic review. *Front Immunol* 11, 604591 (2020).

32. Huang, W., Hickson, L. J., Eirin, A., Kirkland, J. L. and Lerman, L. O. Cellular senescence: the good, the bad and the unknown. *Nat Rev Nephrol* 18, 611–27 (2022).

33. Lee, K. A., Flores, R. R., Jang, I. H., Saathoff, A. and Robbins, P. D. Immune senescence, immunosenescence and aging. *Front Aging* 3, 900028 (2022).

34. Memory immune cells are not the only type of immune cell that becomes senescent; others do too. In fact, senescence is

not really just one thing. For example, some types of macrophage cell can change into something best described as a senescence-like state. A build-up of these causes problems when they accumulate in the walls of blood arteries, as happens in atherosclerotic lesions.

35. Shaw, B. E., et al. Development of an unrelated donor selection score predictive of survival after HCT: donor age matters most. *Biol Blood Marrow Transplant* 24, 1049–56 (2018).
36. Budamagunta, V., Foster, T. C. & Zhou, D. Cellular senescence in lymphoid organs and immunosenescence. *Aging* 13, 19920–41 (2021).
37. Hellmich, C., et al. p16INK4A-dependent senescence in the bone marrow niche drives age-related metabolic changes of hematopoietic progenitors. *Blood Advances* 7, 256–68 (2023).
38. Mogilenko, D. A., Shchukina, I. and Artyomov, M. N. Immune ageing at single-cell resolution. *Nature Reviews Immunology* 22, 484–98 (2022).
39. As well as these issues in bone marrow, senescent cells also accumulate in lymph nodes and the spleen, other places where immune cells tend to be present in large numbers. This correlates with changes in these tissues and organs. For example, in the spleen, the discrete regions which are rich in T cells and B cells become less clearly defined as we age.
40. Peeper, D. S. Old cells under attack. *Nature* 479, 186–7 (2011).
41. Baker, D. J., et al. Clearance of p16Ink4a-positive senescent cells delays ageing-associated disorders. *Nature* 479, 232–6 (2011).
42. Baker, D. J., et al. Naturally occurring p16Ink4a-positive cells shorten healthy lifespan. *Nature* 530, 184–9 (2016).
43. Roy, A. L., et al. A blueprint for characterizing senescence. *Cell* 183, 1143–6 (2020).

44. McHugh, D., *et al.* COPI vesicle formation and N-myristoylation are targetable vulnerabilities of senescent cells. *Nature Cell Biology* 25, 1804–20 (2023).
45. van Deursen, J. M. Senolytic therapies for healthy longevity. *Science* 364, 636–7 (2019).
46. Claesson, M. J., *et al.* Gut microbiota composition correlates with diet and health in the elderly. *Nature* 488, 178–84 (2012).
47. Lythe, G., Callard, R. E., Hoare, R. L. and Molina-París, C. How many TCR clonotypes does a body maintain? *Journal of Theoretical Biology* 389, 214–24 (2016).
48. At one time, the thymus was thought to be entirely uninteresting. It contains lots of dead immune cells, and so was assumed as where immune cells went to die. Jacques Miller, while studying for his PhD at the Chester Beatty Research Institute, London, along with many others, overturned this view. Miller found that mice with their thymus removed very early in life were unable to fight off different types of infection. He also found that mice lacking a thymus did not reject transplanted skin from a genetically different animal. From this it was evident that the thymus was not uninteresting but crucial for establishing an effective immune system.
49. I have written about how all this was discovered in my first book, *The Compatibility Gene* (Penguin, 2013).
50. Shanley, D. P., Aw, D., Manley, N. R. and Palmer, D. B. An evolutionary perspective on the mechanisms of immunosenescence. *Trends in Immunology* 30, 374–81 (2009).
51. Palmer, D. B. The effect of age on thymic function. *Front Immunol* 4, 316 (2013).
52. Brodin, P. and Davis, M. M. Human immune system variation. *Nature Reviews Immunology* 17, 21–29 (2017).
53. Cannon, M. J., Schmid, D. S. and Hyde, T. B. Review of cytomegalovirus seroprevalence and demographic characteristics

associated with infection. *Reviews in Medical Virology* 20, 202–13 (2010).
54. Manicklal, S., Emery, V. C., Lazzarotto, T., Boppana, S. B. and Gupta, R. K. The 'Silent' Global Burden of Congenital Cytomegalovirus. *Clinical Microbiology Reviews* 26, 86–102 (2013).
55. Brodin, P., *et al.* Variation in the human immune system is largely driven by non-heritable influences. *Cell* 160, 37–47 (2015).
56. Chidrawar, S., *et al.* Cytomegalovirus-seropositivity has a profound influence on the magnitude of major lymphoid subsets within healthy individuals. *Clinical and Experimental Immunology* 155, 423–32 (2009).
57. Furman, D., *et al.* Cytomegalovirus infection enhances the immune response to influenza. *Science Translational Medicine* 7, 281ra243 (2015).
58. Wang, H., *et al.* Cytomegalovirus infection and relative risk of cardiovascular disease (ischemic heart disease, stroke, and cardiovascular death): a meta-analysis of prospective studies up to 2016. *J Am Heart Assoc* 6, e005025 (2017).
59. Kaczorowski, K. J., *et al.* Continuous immunotypes describe human immune variation and predict diverse responses. *Proc Natl Acad Sci* 114, E6097–E6106 (2017).
60. Young, M., Benjamin, B. and Wallis, C. The mortality of widowers. *Lancet* 282, 454–7 (1963).
61. Boyle, P. J., Feng, Z. and Raab, G. M. Does widowhood increase mortality risk?: testing for selection effects by comparing causes of spousal death. *Epidemiology* 22, 1–5 (2011).
62. Knowles, L. M., Ruiz, J. M. and O'Connor, M.-F. A systematic review of the association between bereavement and biomarkers of immune function. *Psychosomatic Medicine* 81, 415–33 (2019).

63. Schultze-Florey, C. R., *et al.* When grief makes you sick: Bereavement induced systemic inflammation is a question of genotype. *Brain, Behavior, and Immunity* 26, 1066–71 (2012).
64. In detail, this is a genetic variation in the promotor sequence of IL-6. The promotor sequence is a small stretch of DNA, outside of the gene itself, which enables the gene to be turned on and off. So genetic variation in the promotor sequence doesn't affect IL-6 protein itself, but can affect its production.
65. Vedhara, K., *et al.* Chronic stress in elderly carers of dementia patients and antibody response to influenza vaccination. *Lancet* 353, 627–31 (1999).
66. Kiecolt-Glaser, J.K., *et al.* Chronic stress and age-related increases in the proinflammatory cytokine IL-6. *Proc Natl Acad Sci USA* 100, 9090–5 (2003).
67. Irwin, M. R., Pike, J. L., Cole, J. C. and Oxman, M. N. Effects of a behavioral intervention, Tai Chi Chih, on Varicella-Zoster virus specific immunity and health functioning in older adults. *Psychosomatic Medicine* 65, 824–30 (2003).
68. Yang, Y., *et al.* Effects of a traditional Taiji/Qigong curriculum on older adults' immune response to influenza vaccine. *Med Sport Sci* 52, 64–76 (2008).
69. This can be done in so-called human challenges, in which a small number of people are intentionally given an infection in a safe way with healthcare support in a specialised setting, but benefits must be balanced against the health risks to volunteers and other ethical issues, so this is done very rarely – not to test the benefits of t'ai chi, for example, but in testing treatments for Covid-19.
70. Jobin, J., Wrosch, C. and Scheier, M. F. Associations between dispositional optimism and diurnal cortisol in a community sample: When stress is perceived as higher than normal. *Health Psychology* 33, 382–91 (2014).

71. Segerstrom, S. C. Optimism and immunity: Do positive thoughts always lead to positive effects? *Brain, Behavior, and Immunity* 19, 195–200 (2005).
72. Lee, L. O., et al. Optimism is associated with exceptional longevity in 2 epidemiologic cohorts of men and women. *Proc Natl Acad Sci* 116, 18357–62 (2019).
73. Emmons, R. A. and McCullough, M. E. Counting blessings versus burdens: an experimental investigation of gratitude and subjective well-being in daily life. *Journal of Personality and Social Psychology* 84, 377–89 (2003).
74. Lawton, G. Oldest woman ever or impostor? The controversial case of Calment. *New Scientist* (24 April 2019).
75. Hjelmborg, J.v., et al. Genetic influence on human lifespan and longevity. *Human Genetics* 119, 312–21 (2006).
76. Caruso, C., et al. How important are genes to achieve longevity? *International Journal of Molecular Sciences* 23, 5635 (2022).
77. Shmookler Reis, R. J., Bharill, P., Tazearslan, C. and Ayyadevara, S. Extreme-longevity mutations orchestrate silencing of multiple signaling pathways. *Biochimica et Biophysica Acta (BBA) – General Subjects* 1790, 1075–83 (2009).
78. Takahashi, K. and Yamanaka, S. Induction of pluripotent stem cells from mouse embryonic and adult fibroblast cultures by defined factors. *Cell* 126, 663–76 (2006).
79. Gill, D., et al. Multi-omic rejuvenation of human cells by maturation phase transient reprogramming. *eLife* 11, e71624 (2022).

9. In the Pipeline – or, what big new ideas are on the horizon?

1. Fleming, A. On the antibacterial action of cultures of a penicillium, with special reference to their use in the isolation of B. influenzae. 1929. *Bull World Health Organ* 79, 780–90 (2001).

2. Chain, E., et al. Pencillin as a chemotherapeutic agent. *Lancet* 236, 226–28 (1940).
3. Huang, Y., et al. Pangenome analysis provides insight into the evolution of the orange subfamily and a key gene for citric acid accumulation in citrus fruits. *Nature Genetics* 55, 1964–75 (2023).
4. The genes which give fruit their taste have been selected for by humankind over centuries, mixing breeds and selecting cuttings to grow. Genetic analysis has revealed specific details of how this works. For example, increasing or decreasing the level of activity of the *PH4* gene in citrus fruit trees is a key determinant of taste because it directly affects citric acid levels. A low level is found in oranges, while higher levels in lemons and limes give them their tartness.
5. Prakash, A. and Baskaran, R. Acerola, an untapped functional superfruit: a review on latest frontiers. *Journal of Food Science and Technology* 55, 3373–84 (2018).
6. There's a famous thought experiment which seems relevant here, sometimes called Theseus's paradox. If a ship has all of its parts slowly replaced, so that eventually nothing of the original ship remains, is it still the same ship? Similarly, if some natural vitamin has each atom replaced, one-by-one, so that nothing has chemically changed, when does it become something else, or something unnatural?
7. In detail, there are actually two types of whooping cough vaccine. One uses whole bacteria, and another uses specific protein components of the germ. A shift away from whole cell vaccination for whooping cough happened in the 1990s to reduce side-effects.
8. Boylston, A. The origins of inoculation. *Journal of the Royal Society of Medicine* 105, 309–13 (2012).

9. Jurin, J. and Osborne, J. A letter to the learned Dr Caleb Cotesworth, FRS of the College of Physicians, London, and Physician to St Thomas' Hospital; containing a comparison between the danger of the natural small pox, and of that given by inoculation. *Phil. Trans. R. Soc.* 32, 213–27 (1722).
10. Watson, O. J., *et al.* Global impact of the first year of Covid-19 vaccination: a mathematical modelling study. *Lancet Infect Dis* 22, 1293–1302 (2022).
11. Du, L., *et al.* The spike protein of SARS-CoV – a target for vaccine and therapeutic development. *Nat Rev Microbiol* 7, 226–36 (2009).
12. Gilbert, S. and Green, C. *Vaxxers: a pioneering moment in scientific history* (Hodder & Stoughton, London, 2021).
13. Jiang, C., *et al.* Distinct viral reservoirs in individuals with spontaneous control of HIV-1. *Nature* 585, 261–7 (2020).
14. McLaren, P. J. and Carrington, M. The impact of host genetic variation on infection with HIV-1. *Nat Immunol* 16, 577–83 (2015).
15. Li, C., *et al.* Mechanisms of innate and adaptive immunity to the Pfizer-BioNTech BNT162b2 vaccine. *Nat Immunol* 23, 543–55 (2022).
16. Barbier, A. J., Jiang, A. Y., Zhang, P., Wooster, R. and Anderson, D. G. The clinical progress of mRNA vaccines and immunotherapies. *Nat Biotechnol* 40, 840–54 (2022).
17. Karikó, K., Buckstein, M., Ni, H. and Weissman, D. Suppression of RNA recognition by Toll-like Receptors: the impact of nucleoside modification and the evolutionary origin of RNA. *Immunity* 23, 165–75 (2005).
18. Karikó, K. *Breaking Through: My Life in Science* (Bodley Head, London, 2023).

19. Corbett, K. S., *et al.* SARS-CoV-2 mRNA vaccine design enabled by prototype pathogen preparedness. *Nature* 586, 567–71 (2020).
20. Kobiyama, K. and Ishii, K. J. Making innate sense of mRNA vaccine adjuvanticity. *Nat Immunol* 23, 474–6 (2022).
21. Xie, C., Yao, R. and Xia, X. The advances of adjuvants in mRNA vaccines. *npj Vaccines* 8, 162 (2023).
22. Rerks-Ngarm, S., *et al.* Vaccination with ALVAC and AIDSVAX to prevent HIV-1 infection in Thailand. *New England Journal of Medicine* 361, 2209–20 (2009).
23. Chen, J., *et al.* The reservoir of latent HIV. *Front Cell Infect Microbiol* 12, 945956 (2022).
24. Cohen, G. B., *et al.* The selective downregulation of class I major histocompatibility complex proteins by HIV-1 protects HIV-infected cells from NK cells. *Immunity* 10, 661–71 (1999).
25. Cai, E., *et al.* Visualizing dynamic microvillar search and stabilization during ligand detection by T cells. *Science* 356, eaal3118 (2017).
26. Kumar, S., Shah, Z. and Garfield, S. Causes of vaccine hesitancy in adults for the influenza and Covid-19 vaccines: a systematic literature review. *Vaccines* 10, 1518 (2022).
27. Vaccine hesitancy comes from many sources: culture, personal beliefs, sometimes including religious beliefs, concerns over safety or a sense that the need isn't there. For some, mistrust of pharmaceutical companies also plays a major role.
28. Olotu, A., *et al.* Seven-Year efficacy of RTS,S/AS01 malaria vaccine among young African children. *New England Journal of Medicine* 374, 2519–29 (2016).
29. Zavala, F. RTS,S: the first malaria vaccine. *J Clin Invest* 132 (2022).

30. Datoo, M. S., *et al.* Safety and efficacy of malaria vaccine candidate R21/Matrix-M in African children: a multicentre, double-blind, randomised, phase 3 trial. *Lancet* 403, 533–44 (2024).
31. Simoni, A., *et al.* A male-biased sex-distorter gene drive for the human malaria vector *Anopheles gambiae. Nature Biotechnology* 38, 1054–60 (2020).
32. Kanizsai, A., *et al.* Fever after Vaccination against SARS-CoV-2 with mRNA-based vaccine associated with higher antibody levels during 6 months' follow-up. *Vaccines* 10, 447 (2022).
33. Aliahmad, P., Miyake-Stoner, S. J., Geall, A. J. and Wang, N. S. Next generation self-replicating RNA vectors for vaccines and immunotherapies. *Cancer Gene Therapy* 30, 785–93 (2023).
34. Crow, J. M. HPV: The global burden. *Nature* 488, S2–S3 (2012).
35. Akhatova, A., *et al.* Prophylactic human papillomavirus vaccination: from the origin to the current state. *Vaccines* 10, 1912 (2022).
36. Karikó, K. *Breaking Through: My Life in Science* (Bodley Head, London, 2023), p. 284
37. Franken, M. G., *et al.* Trends in survival and costs in metastatic melanoma in the era of novel targeted and immunotherapeutic drugs. *ESMO Open* 6, 100320 (2021).
38. Carvalho, T. Personalized anti-cancer vaccine combining mRNA and immunotherapy tested in melanoma trial. *Nat Med* 29, 2379–80 (2023).
39. Weber, J. S., *et al.* Individualised neoantigen therapy mRNA-4157 (V940) plus pembrolizumab versus pembrolizumab monotherapy in resected melanoma (KEYNOTE-942): a randomised, phase 2b study. *Lancet* 403, 632–44 (2024).

40. Ott, P. A., *et al.* A Phase Ib trial of personalized neoantigen therapy plus anti-PD-1 in patients with advanced melanoma, non-small cell lung cancer, or bladder cancer. *Cell* 183, 347–62 e324 (2020).
41. Carlton, L. H., McGregor, R. and Moreland, N. J. Human antibody profiling technologies for autoimmune disease. *Immunologic Research* 71, 516–27 (2023).
42. Some of the ideas in this section also appeared in an article I wrote for *New Scientist* magazine, published in June 2021, 'Engineered immunity: Redesigning antibodies to better fight disease'.
43. Wu, L., *et al.* Trispecific antibodies enhance the therapeutic efficacy of tumor-directed T cells through T cell receptor co-stimulation. *Nature Cancer* 1, 86–98 (2020).
44. Gauthier, L., *et al.* Multifunctional natural killer cell engagers targeting NKp46 trigger protective tumor immunity. *Cell* 177, 1701–13 e1716 (2019).
45. Demaria, O., *et al.* A tetraspecific engager armed with a non-alpha IL-2 variant harnesses natural killer cells against B cell non-Hodgkin lymphoma. *Science Immunology* 9, eadp3720 (2024).
46. Xu, L., *et al.* Trispecific broadly neutralizing HIV antibodies mediate potent SHIV protection in macaques. *Science* 358, 85–90 (2017).
47. Promsote, W., *et al.* Trispecific antibody targeting HIV-1 and T cells activates and eliminates latently-infected cells in HIV/SHIV infections. *Nature Communications* 14, 3719 (2023).
48. Cartwright, A. N., Griggs, J. and Davis, D. M. The immune synapse clears and excludes molecules above a size threshold. *Nature Communications* 5, 5479 (2014).
49. Scully, M., *et al.* Caplacizumab treatment for acquired thrombotic thrombocytopenic purpura. *New England Journal of Medicine* 380, 335–46 (2019).

50. Sigoillot, M., *et al.* Domain-interface dynamics of CFTR revealed by stabilizing nanobodies. *Nature Communications* 10, 2636 (2019).
51. Mitra, A., *et al.* From bench to bedside: the history and progress of CAR T cell therapy. *Front Immunol* 14, 1188049 (2023).

Conclusion: the overarching journey

1. Kahneman, D. *Thinking, Fast and Slow* (Farrar, Straus and Giroux, New York, 2011).
2. Shimomura, O., Shimomura, S. and Brinegar, J. H. *Luminous Pursuit: Jellyfish, GFP, and the unforeseen path to the Nobel Prize* (World Scientific, Hackensack, New Jersey, 2017).

Index

ADHD (attention deficit/
   hyperactivity disorder) 143
adrenaline 85–6, 102
advertising 34–5
ageing 6, 25, 138, 152–72, 199
   'biological age' 160
   and exercise/fitness 93–6
   and genes 170–1
   and inflammation 93, 153–4
   and quality of life 172
   and sleep 120, 130
AIDS 177–8
Akbar, Arne 157
alcohol 31–2
Alexander, Albert 174
allergies 36, 38, 51
   and depression 137–8
   and stress 105–9
Alzheimer's disease 51, 82
Amish children 39
ankylosing spondylitis 11, 122
anorexia 74
antibiotics 37–8, 48, 53, 57, 89
antibodies 111, 189–91, 205
antidepressants 136
anti-histamines 106
anti-inflammatory drugs 149
antioxidants 18, 205
anti-TNF therapy 116, 122,
   133–4, 137, 150

anxiety 136, 138, 142
appendicitis 36–7
appetite 97
arthritis 3, 28, 32, 50–1, 116,
   122, 133–4
Asian population 126
aspirin 149
asthma 38–40, 50, 52, 92, 105–6,
   121–4, 136
AstraZeneca 177
athletes 22, 92
autism 44–7, 148
autoimmune diseases 3, 43, 107,
   133, 164, 205
   and eating disorders 74
   and vitamin D 27–8

babies 120, 147
   premature 53
bacteria 3, 37, 42–3, 50–7, 143,
   175, 205
balance exercises 80
Barker, David 36–7
bereavement 168
biobanks 10
BioNTech 180, 185–6
bipolar disorder 143, 150
Black Death 14
blood pressure 81, 102, 136
blood sugar 61–2

body mass index (BMI) 58
bone marrow 162
bowel cancer 64
bowel disease 45, 53–4, 134
brain injuries 51
BRCA1 gene 10–11
breast cancer 64, 77, 123
breast milk 40, 48
broccoli 31
bulimia 74
bullying 118
butyrate 50–1, 164, 206

Calment, Jeanne 170
Cameron, Ewan 23–4
cancer 10–11, 23–4, 31, 64–70, 77
   and ageing 158–60
   checkpoint inhibitors 43–4, 67, 69
   and exercise 83–8
   new treatments 191–2
   and sleep 124–7
   vaccines for 182, 184–8
   and weight 64–7, 69
Cancer Research UK 86
CAR T cell therapy 192
carbohydrates, simple 30
cardiovascular disease 30, 77, 154
caregiving 168–9
cats 145
cells 206
cervical cancer 184–5
checkpoint inhibitors 206–7
chickenpox 91, 166, 169

children 120
   with allergies 51
   and antibiotics 37–8
   on farms 38–40
   microbiome development 48
cleaning 38
Clostridium difficile (*C. diff.*) 47–8, 55–6, 88
colds 89
   and vitamin C 15, 17–18, 21–3, 32
colitis 44–6, 134
colorectal cancer 125–6
competitiveness 100
correlations 6, 89
cortisol 7, 99–104, 108, 111–13, 118, 122, 169, 198, 207
Covid-19 8, 12–13, 73, 88, 104, 108, 138
   and age 157
   and obesity 70–2
   vaccination 129, 148, 177–8, 180
   and vitamin deficiencies 27
Crohn's disease 134
cystic fibrosis 192
cytokines 115–17, 121, 126, 133, 199, 207
   and ageing 154–7, 158
   and mental health 137–41, 146–8, 150
cytomegalovirus (CMV) 91, 166

death 168
dementia 51, 82, 168
dendritic cells 207

INDEX

Dengue virus 118
depression 44, 82, 134
  comorbid 135–41, 199
  early indicators of 138, 141
  as heritable 146
Descartes, René 133
diabetes
  type 1 3, 61–2
  type 2 30, 60–3, 75, 77, 136–7
diarrhoea 30, 47
diphtheria 178–9
dirt, impact on immune health 36–7
divorce 103
dopamine 143
dust 38–9, 41

E. Coli 50, 88
eating disorders 74
Ebola 138
eczema 38
endometrium 64
Epstein-Barr virus 160
euthanasia 172
exercise 22, 40, 70, 77–96, 198
  and cancer 83–8
  and immune cells in the blood 80–1
  and infections 88–91
  recommended levels 80
  strenuous 91–2
  types of 94

faeces 42
  faecal transplants 55–6
farms 38–41, 83

fascism 18–20
fasting 73
Feldmann, Marc 133
fibre, dietary 36, 50, 55, 57
fight-or-flight response 101, 107
Fleming, Alexander 173–4
flu 91, 103, 138
  and age 92–3, 156
  and obesity 71–2
  vaccination 94–5, 167, 169
foods
  anti-inflammatory 31–2
  marketing 5, 58
  processed 15, 30
Friedman, Meyer 99
fungi 2, 35, 39, 207, 209

genes 10–11, 207–8
  editing 193
Gilbert, Sarah 177
Glenny, Alexander 178–9
GLP-1 hormone 75–6
gratitude 169
green tea 31
gut health 41–2
gyms 77

happiness 201
hay fever 36, 38
healing time 152
heart disease 25
herpes 103
histamine 106, 208
HIV 11, 72, 103, 166, 191
  potential vaccine 177–8, 181–2
  and vitamin C 25

INDEX

Holmes, Eamonn 173
hormones 13, 64, 108, 208
   and sleep 121
   and stress 101–2
hospital stays 156
human papilloma virus (HPV) 179, 184–5
Huntington's disease 44
Hutterite children 39
hydrocortisone cream 108
hygiene hypothesis 37–8, 208

Ibuprofen 149
immune health products 5, 35, 52–3, 56–7, 173
immune therapies 43, 126, 190
infections
   and exercise 88–91
   impact of stress 103–4
   susceptibility to 9–11, 27
'inflammaging' 153
inflammation/inflammatory diseases 30, 44–5, 77, 134, 209
innate immune response 208
insulin 61–3, 208
International Space Station 105
intestine 42, 49
irisin 82
irritable bowel syndrome 143

jellyfish 201
Jenner, Edward 7
Jolie, Angelina 10

Kahneman, Daniel 195, 199
Karikó, Katalin 177, 180, 185–6
killer T cells 209

leukaemia 193
life expectancy 100, 153, 163, 170–1
light 142
liver cancer 64
liver disease 30
loneliness 118
lung cancer 64
lungs 52, 71–2
lupus 3, 122
Lynch, Lydia 65

Maini, Ravinder 133
malaria 182–4
malnutrition 75
measles 30
Mediterranean diet 31
melanoma 44, 69, 186
memory 117
mental health 135–51, 199
   and genes 144–6
   and the gut microbiome 141–3
   and inflammation 149
Merck 186
MERS 138
mice 44–5, 50, 65, 84–5, 90, 123, 139–40
   and Covid-19 treatments 104
microbiome 34, 41–57, 198, 209
   manipulating 54–6
   and mental health 141–3
milk, raw 40–1

mindfulness 110–11
Moderna 180, 186
Moertel, Charles 24
mosquitoes 118
mould 38, 40
mouth microbiome 52
MRI (magnetic resonance imaging) 121, 134
mud baths 40
multiple sclerosis 12–13, 45–6, 51, 60, 123
   and mental health 135, 137
   and vitamin D 28
myokines 82, 93

nanobodies 192
napping 126, 130
narcolepsy 117
natural killer (NK) cells 65–6, 83–6, 108, 190–1, 209
'natural' products 173–4
nature 111, 113
Neanderthals 120
nematode worms 170–2
neurological conditions 44, 51
neutrophils 81
niacin 20
night workers 123–4

obesity 60, 136, 154, 198
   and cancer 64–7, 69
   and Covid-19 70–2
   and diabetes 62–3
   impact on lungs 71–2
oestrogen 64, 121
omega 3 fatty acids 30

optimism 111–12, 169
orange juice 6, 174, 196
osteoarthritis 32
oxygen 122
Ozempic 75

pain 122
pandemics 189 *see also* Covid-19
Parkinson's disease 51, 75, 136
Pauling, Linus 18–25, 196
penicillin 173–4
peppers 31
perfectionism 100
perfumes 131
personality, type A 99–100
pneumonia 30
positive mindset 111–12, 169
prebiotics 55
pregnancy 42, 48, 121, 147–8, 166–7
   antibiotics during 37–8
probiotics 35, 52–4, 142–3, 175
progesterone 121
prostate cancer 77
protein molecules 210
psoriasis 28–9, 122, 134
puberty 42

Ramon, Gaston 178
relaxation techniques 110–11
renal cancer 69, 77
resistance training 94
rheumatoid arthritis 3, 28, 50–1, 60, 116, 122, 133–5, 149
   and ageing 152, 154–5
   and alcohol 32

INDEX

RNA vaccines 184
Rosenman, Ray 99
running 80–1, 84, 90, 92

Şahin, Uğur Şahin 186
SARS 138
saturated fats 30
schizophrenia 44, 144, 144–6
sedentary lifestyle 79–80
Selye, Hans 97–8, 100, 112–13
senescence 157–64
sepsis 154, 210
sex differences 12, 107, 128–9
shift workers 123–4
Shimomura, Osamu 201
shingles 91, 166
sitting 79–80
skin 52, 152
sleep 60, 92, 97, 115–32, 126, 169, 199
   apnoea 122, 130
   and brain structure 121
   disrupted 122, 124
   environment 126
   how to sleep well 130–1
   impact of odours on 131
   impact on illness symptoms 121–2
   impact on vaccines 127–9
   lack of 117, 123–4
   optimal amount 119–20, 130
   personalised analysis 131
   reasons for 118
   regulating 116–18
smallpox 7, 176

smoking 28, 112–13, 159
snake venom 192
snoring 60
space travel 105–6
Spanish Inquisition 15
spinal cord injuries 51
stem cells 63, 162, 171
Stone, Irwin 21
Strachan, David 36–7, 41, 57
stress 2, 7, 97–114, 172, 198
   body's responses to 101–4
   causes 98–9
   and sleep 130
strokes 51, 60, 77
sugar 30, 58
sunlight 158
Szent-Györgyi, Albert 21

T regs 50–4, 67, 93, 209
t'ai chi 110, 113, 169
technology 100, 171
telomeres 158–60
thermal water baths 40
thinking 199–201
thrombocytopenic purpura (TTP) 192
thymus 165–6
tiredness 116–18, 121, 126
Toxoplasma gondii 145
trans men 13
tuberculosis 30
tuberculosis (TB) 72–3
tumour necrosis factor (TNF) 210
turmeric 31
twins 15, 45, 170

ulcers 98
ultramarathons 92
ultra-processed foods 15
underweight 73
urinary tract infections 89

vaccination 7–8, 12, 73, 93, 103, 148, 173
  for cancer 182, 184–8
  development of 175–88
  for HIV 177–8, 181–2
  mRNA vaccines 177, 179–80, 185
  personalised vaccines 186–7, 199
  responses to 94–5, 167
  self-replicating RNA vaccines 184
  and sleep 127–9
vegetables, anti-inflammatory 31
viruses 160, 210
vitamin A 30, 40, 198, 210
vitamin B_6 46–7
vitamin C 6–7, 15, 17–18, 32, 174–5, 210
  and cancer 23–5
  high doses 21–3, 196, 198
  and HIV 25
vitamin D 26–9, 32, 40, 198, 210
  and autoimmune diseases 27–8
  supplements 29

walking 111, 113
water, clean 38
Wegovy 75
weight 198
  and ageing 154
  and cancer 64–7, 69
  and Covid-19 70–2
  and diabetes 62–3
  impact on immune health 58–76
  impact on lungs 71–2
  losing 74–5
  and mental health 136
  treating with drugs 75–6
Weissman, Drew 177, 180
The West Wing 99
Western diet 30
whooping cough 175
Willenberg, Loreen 178
womb lining 64
women 12, 107, 128, 138, 159

X-Men 14

Yamanaka, Shinya 171
yoga 110
yoghurt 34–5, 52–3, 175

Zelenay, Santiago 86
Zika virus 118

About the Author

Daniel M. Davis is head of Life Sciences and professor of Immunology at Imperial College London. His previous books include *The Beautiful Cure*, which was shortlisted for the Royal Society Science Book Prize, and *The Secret Body*, which was described variously as 'an inspiration' by Tim Spector, 'beautifully rendered' by Brian Cox and 'masterful' by Alice Roberts. Davis has published over 150 research papers and is a fellow of the Academy of Medical Sciences. In 2025, he was awarded an MBE for services to science communication.